T0245265

CAMBRIDGE LIBRARY COLLECTION

Books of enduring scholarly value

Life Sciences

Until the nineteenth century, the various subjects now known as the life sciences were regarded either as arcane studies which had little impact on ordinary daily life, or as a genteel hobby for the leisured classes. The increasing academic rigour and systematisation brought to the study of botany, zoology and other disciplines, and their adoption in university curricula, are reflected in the books reissued in this series.

Historical and Biographical Sketches of the Progress of Botany in England

Richard Pulteney (1730–1801) was a Leicestershire physician whose medical career suffered both from a lack of aristocratic patronage and from his dissenting religious background. However, his lifelong interest in botany and natural history, and particularly his work on the new Linnaean system of botanical classification, led to publications in the *Gentleman's Magazine* and the *Philosophical Transactions of the Royal Society*. He was elected a Fellow of the Royal Society in 1762. His book on Linnaeus (also reissued in this series), first published in 1782, was later considered to be of great significance for the acceptance in England of the Linnaean system, and this two-volume work, published in 1790, is still relevant to the study of the history of botany. Volume 2 includes the development of botanical gardens, famous figures such as Dillenius and Sherard, and the study of botany in Scotland and Ireland.

Cambridge University Press has long been a pioneer in the reissuing of out-of-print titles from its own backlist, producing digital reprints of books that are still sought after by scholars and students but could not be reprinted economically using traditional technology. The Cambridge Library Collection extends this activity to a wider range of books which are still of importance to researchers and professionals, either for the source material they contain, or as landmarks in the history of their academic discipline.

Drawing from the world-renowned collections in the Cambridge University Library, and guided by the advice of experts in each subject area, Cambridge University Press is using state-of-the-art scanning machines in its own Printing House to capture the content of each book selected for inclusion. The files are processed to give a consistently clear, crisp image, and the books finished to the high quality standard for which the Press is recognised around the world. The latest print-on-demand technology ensures that the books will remain available indefinitely, and that orders for single or multiple copies can quickly be supplied.

The Cambridge Library Collection will bring back to life books of enduring scholarly value (including out-of-copyright works originally issued by other publishers) across a wide range of disciplines in the humanities and social sciences and in science and technology.

Historical and Biographical Sketches of the Progress of Botany in England

VOLUME 2

RICHARD PULTENEY

CAMBRIDGE UNIVERSITY PRESS

Cambridge, New York, Melbourne, Madrid, Cape Town,
Singapore, São Paolo, Delhi, Tokyo, Mexico City

Published in the United States of America by Cambridge University Press, New York

www.cambridge.org
Information on this title: www.cambridge.org/9781108037334

© in this compilation Cambridge University Press 2011

This edition first published 1790
This digitally printed version 2011

ISBN 978-1-108-03733-4 Paperback

HISTORICAL AND BIOGRAPHICAL

SKETCHES

OF THE PROGRESS OF

BOTANY

IN ENGLAND,

FROM

ITS ORIGIN

TO THE

INTRODUCTION OF THE *LINNÆAN* SYSTEM.

BY

RICHARD PULTENEY, *M.D. F.R.S.*

IN TWO VOLUMES.

———————

VOL. II.

———————

LONDON:

PRINTED FOR T. CADELL, IN THE STRAND.

1790.

T O

SIR GEORGE BAKER, BART.

PRESIDENT OF THE ROYAL COLLEGE
OF PHYSICIANS,

PHYSICIAN TO THEIR MAJESTIES,

FELLOW OF THE ROYAL SOCIETY,

AND OF THE SOCIETY OF ANTIQUARIES,
&c. &c. &c.

As eminent for thofe Endowments which dignify the
Characters he fo honourably fupports, as for that
Learning and Science which have moft
defervedly raifed him to the
Attainment of them:

A N D,

T O

MAXWELL GARTHSHORE, *M.D.*

FELLOW OF THE ROYAL COLLEGE OF
PHYSICIANS, EDINBURGH,

OF THE ROYAL SOCIETY,

AND OF THE SOCIETY OF ANTIQUARIES,

AND PHYSICIAN EXTRAORDINARY
TO THE BRITISH LYING-IN
HOSPITAL, &c.

Equally eftimable for Learning and Skill in the Art of
Medicine, as for that Philanthropy which endears
him to all his Friends and Acquaintance:

AS

AS A TRIBUTE

OF THE MOST UNFEIGNED

RESPECT AND ESTEEM;

AND AS A GRATEFUL MEMORIAL

OF THAT UNINTERRUPTED FRIENDSHIP

WITH WHICH BOTH HAVE

LONG HONOURED HIM,

THIS VOLUME IS INSCRIBED,

BY THEIR MOST FAITHFUL

AND OBEDIENT,

HUMBLE SERVANT,

RICHARD PULTENEY.

BLANDFORD,
Feb. 28, 1790.

TABLE of CHAPTERS

IN VOLUME II.

Chap. 34.

TABLE OF CHAPTERS.

TABLE OF CHAPTERS.

VOL. II.

Errors in the Printing.

Page 64. line   1. *for* the *read* a.
— 66. —   6. and 7. *dele the inverted Commas.*
— 92. — 23. *for* LHWYD, *read* LLHWYD.
— 200. —   8. — Mackenbay, *r.* Mackenboy.
— 250. — 15. — LINÆUS, *r.* LINNÆUS.
— 338. — 15. — the *r.* a.
— 345. — 20. *after* HANS *add* SLOANE.
— 348. — 21. — 1754, — Dr. J. GRUF-
                                    BERG.

HISTORICAL AND BIOGRAPHICAL

SKETCHES

OF THE

PROGRESS OF BOTANY,

IN ENGLAND.

CHAP. 27.

Earlieſt notices of botany in Scotland—Alan Ogil-
by—*Dr.* Cargill ; *the correſpondent of* Bauhine
and Lobel—*The* Balfours—Sibbald, *Anecdotes
of—His* Prodromus Hiſtoriæ Naturalis Scotiæ
—Cor-meille—*Hiſtory of* Fife *and* Kinroſs—
His other writings.
Wallace —Preſton —Alſton, *Memoirs of*—Index
Officinalium — Tirocinium — *Adverſe to the*
Linnæan *ſyſtem*—Materia Medica.

SIBBALD.

IT was late before natural hiſtory aroſe in
Scotland. The ſtory of a king *Joſina,*
who is chronicled to have lived more than
150 years before the Chriſtian æra, having
written a book *De Viribus Herbarum,* is
not worth a comment. *Fingal* is ſaid to

have been well acquainted with the virtues of herbs : and *Temory* healed the wounds of his countrymen, by his skill in vulnerary vegetables.

Alan O G I L B Y, who flourished about 1471, a native of *Scotland*, after having travelled through the east, and resided some time at *Conftantinople*, fixed at *Venice*. Besides his eminent acquaintance with the oriental languages, he is celebrated for his knowledge of natural history. He left a book *De Balneis*, and six books *De Virtutibus Herbarum*.

Of Dr. *James* C A R G I L L, of *Aberdeen*, I can produce no material anecdotes, although he merits particular remembrance; since it is manifest, from the nature of his communications to his friends, both on the continent, and at home, that he must have been extremely well acquainted with the botany of the age. There is sufficient evidence that he had studied botany and anatomy at *Bafil*, during the time that *Cafpar* B A U H I N E held the professorship in those sciences, for whom a chair was first erected in that city, in 1589. This celebrated professor enumerates Dr. C A R G I L L among those who

transmitted

tranfmitted feeds and fpecimens to him. GESNER records the fame fervices on his part. At home, LOBEL, in his " *Adver-faria*," acknowledges the like communications, and repeatedly fpeaks of him in very refpectable terms, as a philofopher, and as well fkilled in the fciences of botany and anatomy. He appears to have been living in the year 1603 ; at which time he fent to *Cafpar* BAUHINE fpecimens of the *Fucus digitatus*, with the defcription, which is feen in the " *Prodromus*" of that author. I know not of any publication from Dr. CARGILL, neither am I acquainted with any fuccefsful efforts in the way of natural hiftory, before the time of the BALFOURS.

The founding of the Botanical Garden and the *Mufeum* at *Edinburgh*, by Sir *Andrew* BALFOUR, may be confidered as the introduction of natural hiftory into *Scotland*. Sir *Robert* SIBBALD, the friend and colleague of Sir *Andrew* BALFOUR, and who himfelf added to the ftores of the *Mufeum*, has written " *Memoria Balfouriana*," purpofely to commemorate the liberal benefactions and encouragements given to

B 2   literature,

literature, by Sir *Jacob* and Sir *Andrew*
BALFOUR.

The Garden was eftablifhed about the
year 1680; and, in 1683, was fo fuccefs-
fully cultivated by *James* SUTHERLAND,
the intendant, that it is faid to have contained 3000 fpecies of plants, difpofed according to MORISON's method. An account of it was publifhed under the title
of " HORTUS MEDICUS EDINBURGEN-
" SIS; or, a Catalogue of the Plants in the
" Phyfic Garden at *Edinburgh*, containing
" their moft proper *Latin* and *Englifh*
" names." By *James* SUTHERLAND. 8°.
pp. 367. Varieties, however, occupy a
large fhare of this Catalogue, and very few
of the native plants of *Scotland* are found
in it. It was to Sir *Robert* SIBBALD that
the firft attempts towards indigenous botany were owing.

Robert SIBBALD was a fellow of the
College of Phyficians at *Edinburgh*, and
the firft medical profeffor inftituted in that
univerfity, about the year 1685. He was
knighted by *Charles* II. and had alfo the
title of king's phyfician and geographer
§.   royal

royal conferred upon him, and was a man of very confiderable and various learning. To the knowledge of his profeffion, he added that of natural hiftory, and antiquities. He was, if not the firft, among the earlieft, who wrote on the antiquities of his country, on which he publifhed feveral learned works, to illuftrate, more efpecially, the hiftory of *Scotland* during the time of the *Romans.*

He publifhed, " Scotia illustrata; *five,* Prodromus Historiæ Natura-lis Scotiæ : *in quo regionis natura, inco-larum ingenia et mores, morbi iifque medendi methodus, et medicina indigena explicantur, et multiplices naturæ partus, in triplici ejus reg-no, vegetabili fcilicet, animali, et minerali ex-plicantur."* 1684, folio; and 1696, folio.

In this volume, which, he tells us, was the work of twenty years, one part is ap-propriated to the indigenous plants of *Scot-land;* it contains obfervations on the medi-cinal and œconomical ufes. A few rare fpecies make their firft appearance in this book, particularly that which Linnæus named *Sibbaldia,* after the author; and the *Ligufticum Scoticum.*

Dr.

Dr. SIBBALD having thrown out some strictures on the mathematical principles of physic, for which the learned Dr. PIT-CAIRN was a strenuous advocate, the latter wrote a severe satire on this work, under the title " *De Legibus Historiæ Naturalis.*" Edin. 1696. But it contains nothing solid, and was thought by some to have been the result of party, if not personal dislike.

Among the " *Miscellanca quædam eruditæ Antiquitatis*" of Sir *Robert*, published in 1710, there is a Dissertation on the *Chara* of CÆSAR *, mentioned also by DIO, on which the soldiers of *Valerius*'s army sub-sisted, under a penury of bread. This root has been by some supposed to be the *Kare-mile*, Carmele, or, as Mr. LIGHTFOOT calls it, the *Corr*, or, *Cor-meille* †, of the *Highlanders*. It is the *Orobus tuberosus Linnæi*, our Wood Pease.

In his " History of the Sheriffdom of " *Fife* and *Kinross*," printed the same year, is a catalogue of plants, chiefly maritime,

---

* *De Bello Civili*, lib. iii. § 40.

† See PENNANT's Tour in Scotland, vol. i. Appendix, 292.

growing

growing about the *Frith of Forth*; among which, he had given to one the name of *Balforiana*, now called *Pulmonaria maritima.*

In the zoological way, Dr. SIBBALD publifhed feparately, " *Phalainologia nova :*" 1692. 4°. or, " Obfervations on fome Ani- " mals of the Whale Genus, lately thrown " on the Shores of *Scotland.*" This tract had merit enough to entitle it to a republication, fo lately as in the year 1773. He meditated a *Cætologia,* together with the hiftory of the other marine animals of *Scotland,* in his fecond volume of the " *Prodromus.*"

In the year 1706, he communicated to the Royal Society an accurate defcription, accompanied with a figure of the animal, and its fhell, named *Balanus Balenæ,* or *Pediculus Ceti* of BOCCONE *(Lepas Diadema* of LINNÆUS, *Syft.* 1108.) Thefe were publifhed in vol. xxv. of the *Philofophical Tranfactions,* p. 2314.

Although Sir *Robert* SIBBALD did not carry his refearches fo far, as to rank high in the character of the naturalift; yet, as

having

having led the way in that branch, and fingularly promoted the ftudy of the antiquities of his country, he is juftly entitled to
that honourable ftation he bears among the
writers of *North-Britain* *.

WALLACE.

In the year 1700, was publifhed, " An
" Account of the Iflands of *Orkney*," by
James WALLACE, M.D. F.R.S. which
contains a catalogue of fome of the indigenous plants of that northern region. *Flora*
is not exuberant in her gifts in the chilling
regions of the north. I have not feen this
book ; but I read, that the *arborefcent*, and
fome other tribes, particularly the *malva-
ceous*, are fparingly feen in thefe iflands.

PRESTON.

I know not whether there was any fuper-
intendant to the Garden of *Edinburgh*, be-

---

* His name was applied by LINNÆUS, in the *Flora
Laponica*, to a fmall plant of the *Pentandrous* clafs; which
was known to *Cafpar* BAUHINE and others, and confidered by them as allied to the *Fragariæ*, and the *Pentaphylla*.
It was firft figured by SIBBALD in his "*Prodromus*;" being found in *Britain* only on the *Highland* mountains.

tween

tween Sutherland, and *George* Preston, whom Blair ſtiles an indefatigable
botaniſt, and who publiſhed, about the year
1710, the following Catalogue, written
in *Latin* and *Engliſh :* " *Catalogus omnium*
Plantarum quas in Seminario Medicinæ dicto
tranſtulit Georgius Prestonus, *Bot. Prof.*
et Hort. Edinburg. Præfectus ex Auctoritate
ejus." 12°. Not having ſeen this volume,
I can give no account of it. A writer of
the ſame name occurs, though I know not
whether the ſame perſon, as a correſpondent of Mr. Ray. See his Letters, p. 308
—316; " Some Obſervations on Mr. Ray's
" *Synopſis,*" by Dr. Preston, tending to
illuſtrate the characters of about fifteen
ſpecies of *Engliſh* plants ; with ſome Strictures on Tournefort's method of claſſi-
fication.

In the year 1716, Mr. *Charles* Alston
ſucceeded Preston as ſuperintendant of
the Garden.

ALSTON.

Charles Alston, as we are informed by
Dr. Hope, was the ſon of Mr. *Alſton,* of
Eddlewood;

Eddlewood; a gentleman of fmall eftate in
the weſt of *Scotland*, and allied to the noble
family of *Hamilton*, who, after having ſtu-
died phyſic, and travelled with ſeveral gentlemen, declined the practice of his profeſſion, and retired to his patrimony. His ſon
Charles was born in the year 1683; and, at
the time of his father's death, was at *Glaſgow*, applying with great aſſiduity to his
ſtudies. On this event, the Ducheſs of *Hamilton* took him under her patronage, and
wiſhed him to have choſen the department
of the law; but his inclination for botany,
and the ſtudy of phyſic, ſuperſeded all other
ſchemes; and, from the year 1716, he entirely devoted himſelf to phyſic.

At the age of thirty-three, he went over
to *Leyden*, to ſtudy under BOERHAAVE,
where he remained near three years. At
that place, he contracted an intimacy with
the late celebrated Dr. *Alexander* MONRO;
and, with him, on their return to *Edinburgh*, projected the revival of medical lectures; where, but little had been done in
that department, ſince the firſt eſtabliſhment of the medical profeſſorſhips in 1665,

under

under Sir *Robert* SIBBALD, and Dr. PIT-
CAIRN. The plan was modelled by that
of *Leyden*. MONRO was appointed to give
lectures in anatomy, and furgery; and AL-
STON in botany, and the *materia medica*.
RUTHERFORD, SINCLAIR, and PLUM-
MER, were foon after appointed to fill up
the other departments: and, to the fpirited
endeavours of thefe celebrated names, the
univerfity of *Edinburgh* owes the rife of
that reputation, which has fince fo defervedly raifed it to be one of the firft fchools
of phyfic in *Europe*.

Dr. ALSTON continued to teach botany,
and the *materia medica*, with unwearied affiduity, until the time of his death, which
took place Nov. 22, 1760, in the 77th
year of his age.

In 1740, Dr. ALSTON publifhed for the
ufe of his pupils, " INDEX PLANTARUM
præcipue OFFICINALIUM, *quæ in Horto Me-
dico Edinburgenfi, Studiofis demonftrantur.*" 8°.

In 1752, " INDEX MEDICAMENTORUM
SIMPLICIUM TRIPLEX." 8°. pp. 172.
1. *Alphabetical*; the officinal names, with
numerous fynonyms, from the beft botanical

cal writers, pp. 118. 2. *Officinal* names only; foffils, vegetables, animals, in the order of his lectures. 3. *Classification* of the officinal names, according to the virtues; beginning with the abforbents, and ending with vulneraries. A table of the dofes of emetics and purgatives.

In botany, Dr. ALSTON's chief perform- ance was, his " *Tirocinium Botanicum Edin- burgenfe.*" 1753. 8° It contains a repub- lication of his " *Index,*" firft printed in 1740; to which he now added the " *Fun- damenta Botanica*" of LINNÆUS. But the bulk of the work is a profeffed attempt to explode the fyftem of the *Swede*, and parti- cularly to invalidate all his arguments for the *fex* of plants. This part of it was tranfla- ted by himfelf, and publifhed the next year in the firft volume of " Effays and Obfer- " vations, phyfical and literary." 8°. Could the doctrine of the *fexes* of plants have been eafily fhaken, the learning and abilities of ALSTON were fufficient to have effected his purpofe. But as it was not at that time fupported by hypothefis alone, fo it has fince gained additional ftrength, by new experiments.

experiments, and found inductions, refult-
ing from them. Nurtured from his early
years in the fyftems of TOURNEFORT,
RAY, and BOERHAAVE, to the firft of
which he had even given improvement, it
is not ftrange, that, at an advanced age,
Dr. ALSTON rejected a fyftem of fo much
novelty, as that of LINNÆUS prefented.
We do not willingly unlearn at fixty,
what has been cherifhed from our earlieft
youth.

Dr. ALSTON's medical papers are, " A
" Differtation on *Tin* as an Anthelmintic ;"
" A Differtation on *Opium* ;" and " A Cafe
" of extravafated Blood in the *Pericardium*."
Thefe are printed in the *Edinburgh Medi-
cal Effays*.

In 1743, he difcovered a property in quick
lime, which led him to believe, that the
power of lime was not exhaufted by repeated affufions of water to the fame lime ; he
adds, even for twenty or thirty times. The
firft notices of this paradox, as he then
called it, were communicated to the *Royal
Society*, and were printed in the forty-
feventh

seventh volume of the *Philosophical Transf-
actions*. This opinion was contested, and
drew him into a controversy with his friend
and colleague, Dr. WHYTT. Having continued his experiments, and enlarged his
obfervations, he published, in 1752, his
" Diflertation on Quick-Lime and Lime
" Water;" republished in 1754, and in
1757; in which he replies to Dr. WHYTT's
Strictures ; and, after enumerating a variety of difeafes, in which lime water has
proved efficacious, confirms the opinion of
his colleague, relating to its lithontriptic
powers.

Dr. ALSTON's Lectures on the *Materia
Medica* were prepared for the prefs before
his deceafe, and were publifhed under the
following title :

" Lectures on the *Materia Medica* ; con-
" taining the Natural Hiftory of Drugs,
" their Virtues and Dofes : alfo, Directions
" for the Study of the *Materia Medica* ; and
" an Appendix on the Method of Prefcri-
" bing. Publifhed from the Manufcript
" of the late Dr. *Charles* ALSTON, Profef-
" for

" for of Botany, and the *Materia Medica*, in
" the Univerſity of *Edinburgh*. By *John*
" Hope, M. D. Profeſſor of Medicine and
" Botany in that Univerſity." In two vol.
4°. 1770. pp. 544 and 584.

The firſt eleven lectures conſiſt of preliminary diſcourſes; on the riſe and progreſs
of this knowledge; on the operation of medicines; of errors concerning the *materia
medica*; on claſſing ſimples according to
their virtues; and ſome account of authors
who have written on ſimples.

In treating on each ſubject, after reciting
the officinal name, and the principal *ſyno-
nyma*, the deſcription, and place of growth,
Dr. Alston gives, in his own words, the
ſenſible qualities, powers, and uſes of each
ſimple; after which follows, in the words
of the authors themſelves, a copious detail
of the opinions of reſpectable writers, relating to each; concluding with a recital of
all the officinal compounds into which each
ſimple enters. Add to this, the reader will
meet with a variety of collateral, and hiſtorical information, which is highly gratifying

ing to all fuch as wifh to extend their enquiries beyond the mere nomenclature, and
quality of each fubftance; and which could
otherwife be acquired only from laborious
refearches.

Although the reader will not find the
author giving implicit belief to the mani-
fold, and vaunted powers, attributed to num-
berlefs fimples, through almoft all preceding writers; but on the contrary, will meet
with judicious doubts, obfervations, and
experiments, yet, Dr. ALSTON's *Materia
Medica* muft be confidered, on the whole,
as exhibiting rather the ftate of it, as it
has been, than as it is, in the works of
LEWIS, BERGIUS, MURRAY, and CUL-
LEN. It is but of late that philofophers
and phyficians have exercifed that degree
of fcepticifm on the power of tmedicines,
which muft ever influence the mind, when
experiments alone form the foundation of
medical practice.

Were it within my plan to extend my
obfervations, I fhould, with grateful pleafure, expatiate on the improved ftate of
botany

botany at *Edinburgh*, after this period; when the zeal, and abilities, of my much-honoured and refpected friend, the late Dr. *John* HOPE, affifted by the royal bounty, enabled him to raife the ftudy of botany to an eminence unrivalled, unlefs at *Upfal*, by any univerfity in *Europe*.

C H A P. 28.

Plukenet—*Short memoirs of—A learned, critical,
and laborious botanist* — *His* Phytographia —
Almageſtum *and* Mantiſſa—*His* Amaltheum—
His works had great merit—Contain near 2800
figures—Plukenet *unmindful of generical characters—His ſtrictures on* Sloane *His works
reprinted in* 1769.
Dr. Uvedale, *of* Enfield.

PLUKENET.

IT has been the fate of many learned
men, who have deſerved highly of the
republic of letters, to have the private circumſtances, and occurrences of their lives,
in a few years, ſo far involved in obſcuri-
ty, that almoſt their immediate poſterity,
howſoever deſirous of gratifying a natural
and laudable curioſity, and of rendering to
their memory that tribute which their ſer-
vices have demanded, have been almoſt
wholly fruſtrated in their endeavours to reſ-
cue them from oblivion.

If

If I miftake not, the truth of this pofition is ftrongly exemplified, in the perfon,
of whom, in the order of time, I am next
to fpeak. Of Dr. *Leonard* PLUKENET, as
far as I can find, there are fcarcely any memorials, but what are to be collected from
the prefaces of his works; and they afford
indeed very fcanty information. He has told
us, that he was born in 1642; but whether he was of *Englifh* origin, and of what
family, does not certainly appear; though it
has been conjectured, that he was of *French*
extraction. Where he received his fcho-
laftic education, or at what place he took
degrees in phyfic, I am not able to afcertain.
Some light would be thrown on this circumftance indeed, by determining, where
his two friends, *William* COURTEN, Efq.
and the Rev. Dr. UVEDALE, of *Enfield*,
were educated; fince he fpeaks of both
thefe gentlemen, as having been his fellow-
pupils : probably it was at *Cambridge*, as
he had a fon, named *Richard*, purfuing his
ftudies in that univerfity, at the publication
of his *Almageftum*, in 1696. Be that as it
may, his writings fufficiently teftify his ex-

tenfive knowledge of the learned languages.
He dates the prefaces to his works, from
Old Palace Yard, *Weftminfter*; where, from
a circumftance mentioned in his *Phytogra-*
phia, it may be inferred, that he had a fmall
garden. I know not that he ever attained
to any confiderable eminence as a practical
phyfician. The contrary may rather be
prefumed, as I do not find his name in feveral lifts of the College of Phyficians,
printed in the firft years of this century:
neither in thofe of the *Royal Society* at the
fame period.

His ardour for his favourite purfuit was
remarkably ftrong; *Ut pene nullus, fic ardeo,*
was his motto. It does not appear, that he
ever had an opportunity of gratifying his tafte
by travelling in fearch of plants. He feems
to have devoted all his leifure to his work
of the *Phytographia*; fparing no pains to
procure fpecimens of rare, and new plants.
He had correfpondents in all parts of the
world; and had accefs to the gardens of
the curious, in the environs of *London*, and
to that at *Hampton Court*, which was in a
flourifhing ftate, from the care which the
king

king and queen beftowed on it. The Earl
of *Portland* alfo, had fo much relifh for
exotics, as to have repeatedly fent *Jacob*
REEDE to the *Weft Indies*, to collect curious productions for the Royal Garden.
PLUKENET was one of thofe to whom Mr.
RAY was indebted for affiftance in the arrangement of the fecond volume of his Hiftory; and that eminent man, every where
bears the ftrongeft teftimony to his merit.
Neverthelefs PLUKENET wanted that patronage, to which his learning, and fcience,
entitled him; and he feems, by his com-
plaints, to have feverely felt it. In the latter part of his life, he appears to have been
at variance with SLOANE and PETIVER;
two of the firft characters of the age, for
knowledge in his own ftudies. He cenfures their writings, it muft be confeffed,
in a ftile of too much afperity. Whether
this alienation from thofe of whom he had
before fpoken in terms of friendfhip, and
refpect, had its origin in jealoufy on the
one hand, or what is more probable, on the
other, in that indignant loftinefs, which
too often accompanies the confcioufnefs of

C 3 neglected

neglected merit; or whether from other
fources, I cannot determine.   It was however probably unfavourable to PLUKENET,
fince SLOANE was at that time rifing faft
into reputation, and influence.   In the mean
while,   no obftacles damped   the zeal of
PLUKENET; he was himfelf at the expence
of his engravings, and printed the work at
his own charge, until the publication of
the laft part, his *Amaltheum*, when he procured a trifling fubfcription from a few of
the nobility, amounting to about fifty-five
guineas.   Towards the clofe of his life, he
is faid to have been affifted by the queen,
and to have obtained the fuperintendency of
the garden at *Hampton Court*, and was honoured with the title of Royal Profeffor of
Botany.

I cannot difcover the exact time of his
deceafe; but it is probable he did not long
furvive his laft publication, in 1705.

There is a copper-plate print of Dr.
PLUKENET, done in the 48th year of his
age, prefixed to the *Phytographia*; with his
arms, field ermine, bearing a bend dexter
engrailed gules.

PLUKENET

PLUKENET had all that enthufiafm, without which, few attain pre-eminence; and as the riches of *Flora* were daily pouring into *Britain*, from all quarters of the globe, he failed not to avail himfelf of every opportunity of adding to his ftores. Indigenous fubjects were, equally with exotics, the objects of this induftrious, and learned collector. Hence at length, his *Herbarium* confifted of eight thoufand plants; an aftonifhing number for a private, unopulent individual to collect! Of thefe, the *Phytographia* is to be confidered as the delineation of the new and rare kinds; and the *Almageftum, Mantiffa,* and *Amaltheum,* as the catalogue of the whole.

The *Phytographia* was publifhed at different times. The firft part under the following title:

" PHYTOGRAPHIA; *five,* STIRPIUM ILLUSTRIORUM *et* MINUS COGNITORUM ICONES." 1691. 4°. *tab.* 1—72.

Pars II. 1691. 4°. *tab.* 73—120.

Pars III. 1692. 4°. *tab.* 121—250.

Pars IV. 1696. 4°. *tab.* 122—328.

Thefe four parts, which conftitute the firft

volume of his works, confift entirely of figures.

In the fame year with the fourth part of the *Phytographia*, came out,

" ALMAGESTUM BOTANICUM; *five, Phytographiæ Plukenetianæ Onomafticon, Methodo Syntheticâ digeftum; exhibens Stirpium exoticarum, rariorum, novarumque Nomina, quæ Defcriptionis Locum fupplere poffint."* 4°. 1696. pp. 402.

PLUKENET follows no fyftem; the Catalogue is alphabetical, and contains near 6000 fpecies, of which he tells us 500 were new. Synonyms are added to each, and references made to thofe figured in the *Phytographia*. No man after *Cafpar* BAUHINE had till then examined the antient authors, with fo much attention, as PLUKENET, in order to fettle the fynonyms with truth: and many critical notes interfperfed, prove his intimate acquaintance with all the refources of botanical literature.

Not folicitous to form new *genera*, he refers, from the conformity of *habit* in almoft all inftances, his new plants to the *genera* of former authors; and, more anxious concerning

concerning the *fpecies*, he has defcribed
them with an accuracy that has been ap-
plauded. Not that PLUKENET was unacquainted with *fyftem*, as is manifeft from
one of his criticifms on SLOANE, in the
Mantiffa, p. 113; and from his Obfervations on the firft edition of Mr. RAY's *Synopfis*, publifhed in the Collection of RAY's
Letters, p. 226.

Four years after the publication of the
PHYTOGRAPHIA, came out, with a continuation of the plates, "ALMAGESTI Bo-
TANICI MANTISSA, *Plantarum noviffime
delectarum ultra Millenarium Numerum complectens.*" 1700. 4°. pp. 192. tab. 329—354.
Befides many new plants, this volume contains very numerous additions to the fynonyms of the *Almageftum*. Many curious critical obfervations, on fome of the plants of
the ancient authors, occur in this volume ;
which evince the depth of his knowledge,
and the extreme pains he took in the inveftigation of his fubjects *. A very copious index to both volumes concludes the work.

It

* See his Obfervations on the *Cedrus*, p. 41 ; on the
Juniper of the Hebrews, p. 109 ; on the *Kinfa* of the
9   Chinefe,

It is in the *Mantiſſa* we firſt meet with ſtrictures on SLOANE and PETIVER. He cenſures PETIVER eſpecially, with a degree of ſatyrical acrimony, for errors in the application of ſynomyms in his *Centuriæ*; and SLOANE for the like miſtakes in his " Ca-" talogue of *Jamaica* plants ;" accuſing the latter of having alſo applied his ſynonyms from the *Phytographia,* without acknowledgments, or any reference. *Hinc illæ lachrymæ!*

Five years after the MANTISSA, he publiſhed his laſt work, " The AMALTHEUM BOTANICUM; ſ. *Stirpium Indicarum alterum Copiæ Cornu, Millenas ad minimam, et bis Centum diverſas Species novas et indictas nominatim comprehendens : quarum ſexcenæ et inſuper ſelectis Iconibus æneiſque Tabulis illuſtrantur.*" 1705. 4°. pp. 216. tab. 351 —454. Some of the tables of this volume belong to the plants of the *Mantiſſa.* It abounds with new ſubjects, ſent from *China* and the *Eaſt Indies,* by Mr. CUNNINGHAM and Mr. BROWN, and with ſome from *Florida.*

Chineſe, or the *Poco ſempie*, p. 111 ; on the *Myrobalans*, p. 132 ; on the *Ginſeng*, p. 135, &c. &c.

PLUKENET'S

PLUKENET's work contains upwards of 2740 figures. Moſt of them were engraved from dried ſpecimens, and many from ſmall ſprigs, deſtitute of flowers, or any parts of fructification, and conſequently not to be aſcertained : ſeveral of theſe, neverthelefs, as better ſpecimens came to hand, are figured a ſecond time, in the ſubſequent plates. As he employed a variety of artiſts, they are very unequally done : thoſe by *Vander Gucht* have uſually the preference. The imperfections of this work, however, are, in a great degree, thoſe of the times ; yet it cannot but be regretted that PLUKENET had it not in his power to have given his figures on a larger ſcale. There are unqueſtionably many varieties exhibited as real ſpecies ; and one great defect runs nearly through the whole work, that the new plants are no further deſcribed, than by the ſpecifical definitions, which, under the want of true generical characters, were then inſufficient.

It is, notwithſtanding, a large magazine of botanical ſtores ; inaſmuch as, no work before publiſhed by one man, ever exhibited ſo great a number of new plants. And as
many

many of the *Englifh* fpecies are here figured, for the firft time, it has been equally acceptable to the lovers of indigenous, as of exotic botany.

LINNÆUS, and others, mention a new edition of PLUKENET's works in 1720. But this was nothing more than the ufual artifice of the bookfeller; who, having purchafed the remaining copies, placed a new title-page. They were, however, reprinted, and divided into four volumes, in 1769, with the addition of a few plates, that were wanting in fome copies of the fourth part of the *Phytographia*. Thofe who occafionally confult this author, will regret, that this opportunity had not been taken, of inferting the additions from the *Mantiffa* into the *Almageftum*, by introducing them in a fmaller character, and placing the pages for both in the margin. The *Herbarium* of PLUKENET came into Sir *Hans* SLOANE's poffeffion, and is now in the *Britifh Mufeum*.

In 1779, an *Index Linnæanus* to the tables was publifhed by Dr. GISEKE, profeffor of poetry, and natural philofophy, in the *Gym-nafium*

nafium of *Hamburgh*, which contains a few notes from a MS. left by PLUKENET *

Dr. PLUKENET has not failed to record the names of a numerous set of benefactors, by whose communications he was, from time to time, enabled to amplify his collection, and introduce many new plants to the knowledge of the curious. Among several others, we find, repeatedly, the names of PETIVER, COURTEN, SHERARD, DU BOIS, Bishop COMPTON, Dr. *Tancred* ROBINSON, Dr. SLOANE, CUNNINGHAM, and UVEDALE. Some of these I shall have occasion to commemorate in separate articles; but I regret that I cannot collect any material anecdotes relating to his friend and fellow collegian, ⸻ UVEDALE, LL. D. of whom PLUKENET ever speaks in a stile which indicates that he held him in great esteem.

* Father PLUMIER complimented this learned botanist, by calling after his name a climbing ivy-leaved plant, of the *Monoecious* class, with a *Monadelphous* flower, described only by himself, and by RUMPHIUS, being a native of both Indies.

UVEDALE.

UVEDALE.

Dr. UVEDALE lived at *Enfield*, where he cultivated a garden, which appears to have been rich in exotic productions. And although he is not known among those who advanced the indigenous botany of *Britain*, yet his merit as a botanist, or his patronage of the science at large, was confiderable enough to incline PETIVER to apply his name to a new plant, which MILLER retained in his Dictionary; but which has fince paffed into the genus *Polymnia*, of the *Linnæan* fyftem; the author of which has neverthelefs retained *Uvedalia*, as the trivial epithet.

CHAP. 29.

Petiver — *Anecdotes of* — *Succefsful in collecting a mufeum of natural curiofities* — *His works* — Centuriæ — Gazophylacium — Middlefex *plants*—Plantæ Chinenfes—Switzerland *plants* Pterigraphia—Englifh Herbal—*Various other lifts—and papers in the* Philofophical Tranf-actions.

P E T I V E R.

CONTEMPORARY with PLUKENET lived Mr. *James* PETIVER, of whom too little intelligence is remaining.

It appears that he was apprenticed to Mr. *Feltham*, apothecary to *St. Bartholomew's Hofpital.* He entered into bufinefs for himfelf in *Alderfgate Street*, where he lived the remainder of his days. He became apothecary to the *Charter Houfe*, and obtained a confiderable fhare of practice in his profeffion.

He had an early propenfity to thefe ftudies, and, excepting Mr. COURTEN, and Dr. SLOANE, feems to have been the *only* one, after

ter the TRADESCANTS, who made any considerable collection in natural history. PETIVER engaged the captains, and surgeons of ships, to bring home specimens, and seeds of plants, birds, stuffed animals, and insects; and he directed their choice, and enabled them to judge, in some measure, of proper objects, by distributing printed lists and directions among them. He was not less anxious to procure, what his native country afforded, and was so successful in his efforts, that Sir *Hans* SLOANE, who afterwards purchased it, offered PETIVER four thousand pounds for his *Museum*, some time before his death: which offer, although it may be considered as a proof of the opulence of Sir *Hans*, is equally so of the extent of the collection.

The allurement of such uncommon curiosities as Mr. PETIVER exhibited, soon obtained him considerable distinction, and his name became well known, both at home and abroad. He was elected into the *Royal Society*; and as his particular attachment was to plants, he became early the correspondent of Mr. RAY, who acknowledges

his

his affiftance in arranging the fecond volume of his " Hiftory of Plants ;" and elfewhere owns his high obligations to him,
for the extent and freedom of his communications.

In the year 1692, preparatory to the publication of his firft work, PETIVER took a
tour into the midland counties of England.
I recollect, on this occafion, the pleafure I
had in my youth, in feeing the *Lichen ju-
batus* growing on the fpot, where, I believe,
he firft difcovered it, on the higheft rocks
in *Charley Foreft, Leicefterfhire.*

Mr. PETIVER's firft publication was,
" MUSEI PETIVERIANI *Centuriæ decem.*"
1692—1703. 8°. Containing the names,
and fynonyms of various rare animals, foffils, and plants ; among which, feveral curious articles, the produce of *England,* are
here firft exhibited ; particularly fome of
the *Cryptogamous* clafs, in the inveftigation
of which he was very fuccefsful.

" GAZOPHYLACII NATURÆ *et* ARTIS
Decades decem." 1702. tab. 100. fol. A
book of great value at the time of its publication, being the engravings, accompanied with

VOL. II.   D   fhort

fhort defcriptions, of animals of all the orders, vegetables, and foffils : among thefe
are many *American* ferns, plants of the *Alps*,
and from the *Cape of Good Hope* ; all, either
very rarely feen before, or nondefcripts.   It
will retain its value while LINNÆUS's writings are in ufe.

Among the provincial lifts of plants,
printed in Bifhop GIBSON's edition of
CAMDEN in 1695, Mr. PETIVER communicated the *Middlefex* plants.   All the
others were drawn up by Mr. RAY, as was
obferved under his article.

Next to the *Gazophylacium* in the order
of time, although not a diftinct work, was
publifhed, in Mr. RAY's third volume of
his Hiftory of Plants, " PLANTÆ RARIO-
RES CHINENSES, MADRASPATANÆ *et*
AFRICANÆ, *à Jacobo* PETIVERO, *ad Opus
confummandum collatæ: cum ejufdem Catalogo
Plantarum in Hortis fuis ficcis confervatarum,
quæ vel ineditæ, aut hactenus obfcurè defcriptæ
funt : adjicitur Farrago Stirpium Indicarum,
et Americanarum incertæ Sedis.*"

The firft of thefe catalogues amounts to
184 plants : thofe of the *Hortus ficcus*, to
more

more than 800 fpecies : the laft to 75.
Although doubtlefs great numbers of thefe
muft have been varieties only, thefe lifts
will yet remain a lafting teftimony of the
early and extreme diligence of this indefatigable collector.

In 1709, he publifhed, without his name,
" A Catalogue of Plants found on the
" mountains about *Geneva,* the *Jura, La*
" *Dole, Saleve;* with others growing in the
" fields, &c. as obferved by GESNER, the
" BAUHINES, CHABRÆUS, and RAY."

" PTERIGRAPHIA AMERICANA : ICO-
NES *continens plufquam CCCC Filicum vari-
arum Specierum.*" Tab. 20. 1712. fol.
The ferns occupy fixteen of thefe tables.
Among thefe are contained moft of Father
PLUMIER's ferns. The four remaining
tables are of fubmarine productions.

Mr. PETIVER neglected no opportuni-
ties of augmenting the *Englifh Flora.* He
was the firft difcoverer of many *Englifh*
plants, as well as of other natural productions, fome of which he figured in the *Gazophylacium;* but he meditated, and in part
executed, (a work that had not been at-

D 2   tempted

tempted before) a fet of diftinct figures of
Britifh plants. Unfortunately he lived not
to finifh it.

This work, which diftinguifhes PETI-
VER as an auxiliary to *Englifh* botany, bears
the title of " A Catalogue of Mr. RAY's
" *Englifh Herbal*, illuftrated with Figures."
fol. 1713. t. 50; and continued " with the
" four-leaved flowers," t. 51—72. fol. 1715.
Twelve plants are engraved on each plate.
The work ends with the feventeenth clafs.
The figures are little more than outlines, but
they are neat; and though they have the
defect of the old herbals, in being all on a
fimilar fcale, were valuable, and efpecially
as pointing out many of the varieties in the
Synopfis of RAY, particularly among the
Apetalous and *Syngenefious* tribes. A new
impreffion of thefe plates was made under
the infpection of Sir *Hans* SLOANE, in
1732.

Thefe were the moft material works of
PETIVER. His fmaller publications a-
mount to a great number, and are of lefs
importance at this day, as being principally
fhort catalogues and fingle tables of rare
plants,

plants, intended, in many inſtances, as in-
ſtructions to his various correſpondents:

Plantarum Etruriæ rariorum Catalogus.
1715. fol. one ſheet.

*Monſpelii deſideratarum Plantarum Cata-
logus.* 1716. fol. one ſheet.

*Plantarum Italiæ marinarum et Graminum
Icones Nomina,* &c. 1715. fol. one ſheet,
with five plates.

Hortus Peruvianus medicinalis: The South
Sea Herbal of FEUILLF's Medicinal Plants.
1715. with five plates.

GRAMINUM, MUSCORUM, FUNGORUM
SUBMARINORUM *et* BRITANNICORUM,
CONCORDIA. 1716. fol.

*Petiveriana, ſ. Collectanea Naturæ domi
foriſque Auctori communicata.* 1717. fol.

Plantæ Sileſiacæ rariores ac deſideratæ.
1717. fol. a ſingle ſheet.

*Plantarum Ægyptiacarum rariorum Icones:
et aliarum Catalogi duo.* 1717. fol. one
ſheet, with two plates.

Plants engraved in Mr. PETIVER's *Engliſh Herbal.* fol. one ſheet.

Hortus ſiccus Pharmaceuticus.
Directions for gathering Plants.

D 3   Beſides

Befides thefe fmall publications, he put forth, at different times, twenty-eight tables of rare plants; of which nineteen contained *American* plants; four, rare plants from various parts of *Italy*; two, *Auftrian* plants; and one, *Indian* roots and gums.

There are more than twenty papers written by PETIVER, and publifhed, at divers times, in the *Philofophical Tranfactions*, between the years 1697 and 1717.

A Catalogue of fome *Guinea* Plants, with their Nature, Names, and Virtues; fent by the Rev. *John* SMITH, from *Cape Coaft*; with Remarks, by Mr. PETIVER. N° 232. Vol. XIX. p. 627.

An Account of forty-fix Plants, collected by Mr. *Samuel* BROWNE, near *Madras*; with the Synonyms, and critical Obfervations, by Mr. PETIVER. N° 244. Vol. XX. p. 313.

Remarks on fome Animals, Plants, &c. fent by the Rev. Mr. *Hugh* JONES, from *Maryland*. N° 246. Vol. XX. p. 396.

An Account of Part of a Collection of curious Plants and Drugs, collected at *Ma-*
dras

dras by Mr. *Samuel* BROWNE, and prefented to the Royal Society by the Eaft India Company: in eight books, publifhed at
different times; the firft in N° 236, the
laft in N° 299. Vol. XXIII.

Mr. PETIVER was among the firft who
purfued the idea that the virtues of plants
might be determined, in general, by their
agreements in natural charaċters and claffes.
I fay purfued, becaufe the idea had been
fuggefted long before. Even CÆSALPI-
NUS, the inventor of fyftem, hints that the
virtues of plants are pointed out by their
ftrudture, and alliance to each other. Thefe
are his words: *Tandem et facultates, quas
medici maxime quærunt, tanquam proprietates
innotefcunt ex naturarum cognitione: quæ enim
generis focietate junguntur, plerumque et fimi-
les poffident facultates* *

PETIVER's paper bears the following
title; " Some Attempts made to prove,
" that Herbs of the fame Make, or Clafs,
" for the generality, have the like Vertue,
" and Tendency to work the fame Ef-
" feċts." N° 255. Vol. XXI. p. 280.

* Cæfalp. Pref. ad Lib. de Plantis
D 4

As a firſt eſſay, Mr. PETIVER has ſucceſſ-
fully treated his ſubject. It is well known
what uſe LINNÆUS and others have ſince
made of it : and Dr. MURRAY, the pre-
ſent Profeſſor, of *Gottingen*, has choſen this
arrangement for his *Apparatus Medicami-
num.* In BLAIR's " Miſcellaneous Obſer-
" vations," publiſhed in 1718, Mr. PE-
TIVER defends his doctrine, in anſwer to
Dr. BLAIR's doubts on that head.

Remarks on ſome Inſects, ſent by Mr.
BANISTER from *Virginia*, in 1680. N°
270.

An Account of ſome Animals, ſent by
Mr. E. BULKELEY from *Madras.* N° 271,
and 276.

A Deſcription of ſome Shells, from the
Molucca Iſlands. N° 274.

An Account of ſome Animals, ſent to
Mr. PETIVER from the *Philippine* Iſlands,
by *Father* GEMELLI. N° 277.

A Deſcription of ſome Shells, ſent from
Madras by the Rev. Mr. LEWIS to Mr.
PETIVER. N° 282.

A Deſcription of ſome Corals and other
Submarines, from the *Philippine* Iſlands,
ſent

fent by *Father* CAMELLI to Mr. PETIVER.
N° 206.

An Account of fome Shells and Animals, fent from *Carolina* to Mr. PETIVER.
N° 299.

A Catalogue of Foffil-Shells, Metals,
and Minerals, fent by Dr. *John* SCHEUCH-
ZER to Mr. PETIVER. N° 301.

An Account of fome Minerals, petrified
Shells, and other Foffils, from *Berlin.*
N° 302.

An Account of a MS. of *Father* CA-
MELLI's, concerning fome Shells, Mine-
rals, Foffils, and of the Warm Baths of the
Philippine Iflands. N° 311.

An Account of fome Swedifh Minerals,
fent to Mr. PETIVER. N° 337.

As Mr. PETIVER accompanied thefe
communications with remarks, the exhibition of fuch objects, from fo intelligent a
naturalift, in the early period of the *Royal
Society,* when the ftudy of nature was in its
infancy, could not fail to convey much information, and excite a curiofity to purfue
with zeal, one of the chief ends of the inftitution.

In

In N° 313, Mr. PETIVER communicated to the Royal Society the manner of making the *Styrax liquida*, which, he fays, is from the bark of the *Rofa mallas* (the *character* of which is unknown) which grows on the ifland *Cobrofs*, in the Red Sea. If this be the origin of the true, or oriental kind, there is likewife a *Styrax liquida*, from the *Liquidambar* tree of *Virginia*. What is generally met with now, is juftly fufpected to be a mixed artificial compofition.

In N° 232, Mr. PETIVER publifhed, under the title of *Botanicum hortenfe*, an Account of divers rare plants, obferved in feveral curious gardens about *London*, particularly in the Phyfic Garden at *Chelfea*. This was continued, in feven feparate tracts, to N° 346. Vol. XXVII. XXVIII. XXIX.

Mr. PETIVER died at his houfe in *Alderfgate Street*, on the 20th of April, 1718. His body was carried to *Cooke Hall*, where, agreeably to the cuftom of the time, it lay in ftate. The pall was fupported by Sir *Hans* SLOANE, Dr. LEVIT, phyfician to the *Charter Houfe*, and four other phyficians. He left five guineas to Dr. *Brady*, for

for preaching his funeral fermon, and fifty pounds to the charity fchool of *St. Ann's, Alderfgate**.

Many of PETIVER's fmaller pieces having become very fcarce, his works, exclufive of his papers in the " Philofophical Tranfactions," were collected and publifhed in 1764, in two volumes in folio, and one in octavo ; with the addition of fome plates, not in the firft edition.

---

* PETIVER's name was annexed by PLUMIER to one of his new *American genera*, of which a fecond fpecies is common in *Jamaica*, and had been defcribed by SLOANE as belonging to the *Verbena* or *Schrophularia* genus.

C H A P.

CHAP. 30.

Perſonal names given to plants—Antiquity of—
Inſtances in the antients—Among the monks—
and the reſtorers of botany: by Geſner *and*
Matthiolus—*Revived by* Plumier—*His liberal*
and impartial uſe of this privilege.
Anecdotes of Plumier—*Account of his works—*
Deſcription of American *plants—*Nova genera
*—*American *ferns—Five hundred 'of his tables*
purchaſed by Boerhaave—*Publiſhed by* Burman
—His L'Art de Tourner.

PERSONAL NAMES OF GENERA.

PETIVER and PLUKENET, as far as I
can find, were the firſt *Engliſh* writers,
who followed the example of PLUMIER in
giving perſonal names to new *genera*. PE-
TIVER is, however, ſeverely reprehended by
LINNÆUS, for having conferred this honour on ſome who did not merit it : He
obſerves juſtly, that it is due to ſuch alone,
as have ſignalized themſelves in the ſcience.
I take this opportunity to remark the riſe
and

and progreſs of this cuſtom, which appears
to be of high antiquity, ſince there are ex-
amples of it among the antient poets, hiſto-
rians, and phyſicians.

Poetry has conſecrated, in this way, the
names of *Adonis, Daphne, Hyacinthus, Nar-*
ciſſus, and others. PLINY informs us, that
Eupatorium is ſaid to be the cognomen of
MITHRIDATES, who firſt diſcovered the
uſes of that plant. *Gentiana*, we are told,
is derived from GENTIUS, king of *Illyria:*
Lyſimachia, from LYSIMACHUS, king of
Sicily: Telephium, from TELEPHUS, king
of *Myſia: Teucrium*, from TEUCER, king of
Troy: Clymenum, from CLYMENUS: *Arte-*
miſia, from the wife of king MAUSOLUS:
Helenium, from HELENA, the wife of *Me-*
nelaus: Euphorbium, from EUPHORBUS,
phyſician to *Juba* II. king of *Mauritania;*
though SALMASIUS avers, that this name
had been in uſe at a much earlier period.
Many other inſtances might be adduced.

In ſucceeding ages, the devotion of the
monks led them to conſecrate a variety
of plants to the ſaints of the kalendar.
Thus we have the *Herba Sancti* ANTO-

NII,

NII, *Epilobium:* S. CHRISTOPHORI, *Actæa:* S. GERARDI, *Ægopodium:* S. RUPERTI, *Geranium:* S. JACOBI, *Senecio:* S. PETRI, *Parietaria*, &c. &c. *John* BAUHINE wrote a treatise, in 1591, now become very scarce, " *De Plantis à Divis Sanctisve Nomen habentibus.*"

The restorers of botany, in a few instances, revived the practice. GESNER, had he lived to finish his plan in his " History " of Plants," intended to have perpetuated the names of his friends, by monuments of this kind. It appears, that he had requested GUILANDINUS, *John* BAUHINE, KENTMAN, CAMERARIUS, and our celebrated countryman Dr. CAIUS, to select from his new species, such as they chose to bear their names, or allow to him the privilege of adopting what he thought most congruous. By the same kind of tribute we learn, from his letters, that he proposed to record the names of his friends GASSERUS, OCCO, ARETIUS, and several others.

MATTHIOLUS, however, actually restored the usage, by the application of the term *Cortusa* to a new plant, in honour of CORTUSUS,

CORTUSUS, the fucceffor of GUILANDI-
NUS, in the garden of *Padua* ; and CLU-
SIUS followed his example, by calling the
Contrayerva of the fhops *Drakæna*, in honour of Sir *Francis* DRAKE ; from which
time it was fparingly practifed, until after
the invention of fyftem ; and the conftruction of generical characters, at the latter
end of the laft century. TOURNEFORT,
PLUMIER, and PETIVER, led the way, and
have been followed by all fucceeding writers
of note, and by none more than by LIN-
NÆUS himfelf. It may be ftiled the *apo-
theofis* of botanifts ; and LINNÆUS may be
compared to the high prieft, who has thus
immortalized a numerous group of celebrated men.

The practice, however, was feverely cen-
fured by fome of the older botanifts, who
objected, that thefe names, having no connection with the form, nature, habit, or
properties of the plant, conveyed no idea
of diftinction. But the objection, if duly
weighed, is of no force ; fince there is not,
perhaps, a fingle appellation, even among
the beft, of *Greek* etymology, however aptly

4   framed,

framed, which conveys any character of the genus, that might not with equal propriety have been applied to a multitude of others.

The free ufe that PLUMIER made of this privilege, in honouring fo great a number of Englifhmen, does equal credit to his own difcernment, and impartiality, and to the merit of thofe on whom he beftowed this laurel. On this account, I hope it may not be ungrateful to the reader, to digrefs fo far, as to introduce a fhort notice of this liberal-minded foreigner.

PLUMIER.

Charles PLUMIER was born at *Marfeilles*, in the year 1646; and, after a claffical education, devoted himfelf to a religious life; and, at the age of fixteen, entered into the order of the *Minime Friars*. Being early inclined to mechanics and philofophy, he ftudied mathematics, at *Touloufe*, under *Pére Maignan*, a celebrated profeffor of the fcience, and of the fame order. In fome of his leifure hours, he amufed himfelf in the practice of Turning, having been taught theart by his father; and became fo great

a proficient,

a proficient, that he wrote a book on the
fubject. Under *Pére Maignan*, he alfo
learned to make lenfes, mirrors, micro-
fcopes, and other inftruments of philofophy; and early acquired the arts of defigning and painting. He was foon after fent
by his fuperiors to *Rome*, where his clofe
application to his ftudies, and to thefe arts,
together with optics and mathematics, de-
ranged his health. In this fituation, he
fought for relaxation in the ftudy of botany,
under the lectures of *Pére* SERGEANT, in
a convent at *Rome*, and in the inftructions
of *Francis de* ONUPHRIIS, an Italian phyfician, and of *Sylvius* BOCCONE, a Sicilian.
By thefe connections, he was infenfibly led
to devote himfelf to his new ftudy. Being
recalled by his order into *Provence*, he was
placed in a convent near *Hyeres*, and obtained leave of his fuperiors to fearch the
coafts of that country, and the neighbouring Alps, for plants. At this time, he conceived a defign of forming a new *Pinax*, or
" General Hiftory of Vegetables," for which
he had made many drawings, and collected
materials; but his fubfequent deftination

VOL. II.　　　　E　　　　prevented

prevented his making an effectual progress
in this defign. He foon after became acquainted with TOURNEFORT, then on his
botanical tour in the South of *France*; and,
together with GARIDEL, profeffor of botany at *Aix*, accompanied that eminent man
in his refearches.

Thus qualified, and while he was herbo-
rizing on the coaft of *Marfeilles*, he was
chofen as the affociate of SURIAN, to ex-
plore the French fettlements in the Weft
Indies. The French were ftimulated to this
expedition, by the fuccefs of our great naturalift, SLOANE, in *Jamaica*. PLUMIER
acquitted himfelf fo well, that he was twice
fent afterwards, at the King's expence, to
compleat the natural hiftory of the *Antilles*;
and was rewarded with the title of Botanift
to the King, and an increafed penfion each
time. He paffed two years in thofe iflands,
and on the neighbouring continent; but
made *Domingo* his principal refidence. He
made defigns and paintings of many hundred
plants, on a fcale as large as the life; befides numerous figures of birds, fifhes, and
infects.

9   On

On his return from his fecond voyage,
by the intereft of M. Pontchartrain,
he procured the firft fpecimen of his labours
to be publifhed at the *Louvre,* under the
title of " *Defcription des Plantes de l'Amerique.*" Fol. 1695. pp. 94. tab. 108. The
firft fifty of thefe tables reprefent Ferns; the
remainder, divers fpecies of the *Arum* genus;
the *Piper, Paffiflorœ, Rajania, Dolichos,* and
various others. The figures confift of little
more than outlines; but being as large as
the life, and drawn with great accuracy and
freedom, they have a very fine effect. The
defcriptions are in French.

On his return from the third voyage, he
fettled at *Paris,* in the character of his order; and, in 1703, publifhed his " *Nova
Plantarum Genera.*" 4°. In this work,
which is conftructed on the plan of Tournefort's " Inftitutions of Botany," the author defcribes, and figures, the characters of
106 new *genera*; among which are many
of the plants ufed in medicine. It is in
this book he pays the tribute, fo often fpecified, not only to the manes of deceafed
botanifts, but to feveral eminent men then

living:

living : he has, in this way, given appellations to more than fifty *genera,* taken from
the names of botanifts.

In the courfe of thefe pages, thofe generical terms, which have been formed from
the names of *Englifh* botanifts of renown,
thus celebrated by *Father* PLUMIER, are
mentioned under their refpective articles :
but I here collect them into one view :

    Gerardia.   *Plukenetia.*
    Lobelia.   *Rajania.*
    Morifonia.   *Sloanea.*
    Parkinfonia. *Turnera.*
    Petiveria.

In the year 1704, he was prevailed on by
M. FAGON, to undertake a voyage to *Peru,*
to difcover and delineate the *Peruvian Bark*
tree. Nothing but the greateft zeal for
fcience, could induce a man at the age of
fifty-eight, to attempt fo perilous a voyage.
While he was waiting for the fhip, to em-
bark with a new viceroy at Port *St. Mary,*
near *Cadiz,* he was feized with a pleurify,
and died.

Having, before his departure from *Paris,*
prepared for the prefs his " *Traité des Fou-*
              †   *geres*

geres de l'Amerique," it was printed in folio,
in 1705; pp. 146. tab. 172. The text in
French and *Latin.* All the Ferns contained
in the former volume enter again into this;
and, as this likewife was printed at the
King's expence, it has all the advantages in
the execution, that accompanies royal munificence.

We are informed by Dr. LISTER, that
PLUMIER left behind him drawings of
plants, fufficient to make ten volumes, equal
to the firft book above mentioned. They
are faid to have amounted to 1400. Some
of thefe remained in the convent at *Paris*,
to which PLUMIER belonged : others were
in the King's library. From the latter,
BOERHAAVE procured copies of 508 fpecies, done by AUBRIET, under the infpection of VAILLANT. Thefe were publifhed
in 1755—1760, by Profeffor BURMAN, at
Amfterdam, in ten *fafciculi*, accompanied
with 262 plates, on which are engraven
upwards of 400 fpecies : and Dr. BLOCH,
of *Berlin*, has procured many of the fifhes
for his late fplendid work on ichthyology.

PLUMIER

PLUMIER was the author of two differtations; one in the " *Journal des Scavans* " of
1694 ; the other in the " *Journal des Tre-*
voux;" to prove that cochineal was an infect ; a fact doubted by many at that time,
and concerning which his own teftimony
had been controverted. In the fame work
occur fome publications by PLUMIER, on
other zoological fubjects.

His *L'Art de Tourner* was firft publifhed
at *Lyons* in 1701 ; and a fecond time at
Paris, fo lately as in 1749, in folio, with
eighty plates. It is fpoken of as a curious
and fingular work, containing the moft
compleat inftructions relating to that art,
that are to be met with.

CHAP.

C H A P. 31.

Banifter—*communicates plants to* Ray—*Perifhed in* Virginia *by falling from the rocks—His papers in the* Philofophical Tranfactions.

Vernon *and* Kreig—*collect an Herbarium in* Maryland.

Cunningham—*enriched exotic botany, by fending plants from* China *and the* Eaft Indies.

Brown, Samuel, *and* Alexander—Glen—*Contemporary of* Ray—*His Herbarium.*

B A N I S T E R.

*J*OHN BANISTER, is mentioned by Mr. RAY in very high terms, as a man of talents in natural hiftory. He firft made a voyage to the *Eaft Indies*, where he remained fome time ; but was afterwards fixed in *Virginia*. In that country he induftrioufly fought for plants, defcribed them, and himfelf drew the figures of the rare fpecies : he was alfo celebrated for his knowledge of infects, and meditated writing the natural hiftory of *Virginia*, for which, Mr. RAY

obferves

obferves that he was every way qualified. He fent to RAY, in 1680, " A Catalogue " of Plants obferved by him in *Virginia*;" which was publifhed in the fecond volume of RAY's Hiftory, p. 1928.

The world was deprived of much of the fruit of his labours, by his untimely death. BANISTER increafed the lift of martyrs to natural hiftory. In one of his excurfions, in purfuit of his object, he fell from the rocks, and perifhed. His *Herbarium* came into the poffeffion of Sir *Hans* SLOANE, who thought it a confiderable acquifition.

The following papers, written by Mr. BANISTER, were publifhed in the *Philofophical Tranfactions*:

A Catalogue of feveral Curiofities found in *Virginia*. N° 198. p. 667.

Obfervations on the *Mufca Lupus* of MOUFFET, in *Virginia*. They relate to the balancers or poifers, called by LIN- NÆUS *Halteres*, fixed under the wings of the order of *Diptera* among infects. N° 198. p. 670.

On feveral Sorts of Snails obferved in *Virginia*. Ib. p. 672.

A Defcription

A Defcription of the *Piftolochia,* or *Serpentaria Virginiana,* the Snake-root of the fhops *(Ariftolochia Serpentaria,* Lin.) N° 247. p. 467 *

VERNON AND KREIG.

About the fame time with BANISTER, as I conjecture, Mr. *William* VERNON, fellow of St. Peter's College, *Cambridge,* and Dr. *David* KREIG, a German phyfician, led by their genius for botany, made a voyage to *Maryland.* They returned, after having collected an *Herbarium* of feveral hundred new and undefcribed plants. Thefe came into the poffeffion of Sir *Hans* SLOANE, by whofe liberal communication they were inferted in the " Supplement " to RAY's Hiftory. Mr. VERNON is alfo fpoken of by RAY, as not lefs fkilful and affiduous in the purfuit of *Englifh* plants,

---

* Mr. HOUSTON confecrated to BANISTER a genus of *Decandrous* climbing plants, which SLOANE, PLUKENET, and PLUMIER had ranked with the *Acer:* But the diftinction of HOUSTON ftood the teft of the *Linnæan* rules, and is preferved in the fexual fyftem.

and

and of all other branches of natural knowledge.   His difcoveries in the *Cryptogamia*
clafs were numerous.

Of Dr. KREIG, I can give no further
account than that he was of *Saxony*. I conjecture, that after his return from *Maryland*, he retired into his native country. He
was the friend and correfpondent of DALE,
who, in his *Pharmacologia*, introduces his
name in the moft refpectful manner, for
notices communicated by him; and ranks
him among the few eminent men of the time,
who excelled in the knowledge of the *Materia Medica* and Chemiftry.   Dr. KREIG
was not living when DALE publifhed the
third edition of the above-mentioned work,
in 1737.

Dr. KREIG communicated to the Royal
Society, " An Account of Cobalt and the
" Preparation of Smalt and Arfenic," ac-
cording to the procefs ufed at the mines of
Shnecbergh, in *Hermanduria*. It was printed
in the *Philofophical Tranfactions*, N° 293,
Vol. xxiv. p. 1754; with figures of the
Furnaces.

CUNNINGHAM.

CUNNINGHAM.

In the period we are now fpeaking of, but few voyagers poffeffed any confiderable knowledge of nature ; or a fpirit of en-quiry, powerful enough to induce them to facrifice their avocations, from interefted purfuits, to the ftudy of natural hiftory. SLOANE, BANISTER, and a few others, were indeed exceptions; and, in this fketch, it would be injurious to his memory, not to add the name of *James* CUNNINGHAM, to whom RAY, and particularly PLUKE-NET and PETIVER, acknowledge important obligations, for his copious communica-tions of new plants.

The merit of Mr. CUNNINGHAM would juftly demand a more complete gratification of curiofity concerning his life and circum-ftances, than I can fupply. I can only collect, that he went out in 1698, as fur-geon to the factory, eftablifhed by the Eaft India Company at *Emuy,* or *Amoy,* on the coaft of *China* ; and afterwards, made a
fecond

fecond voyage in the fame capacity, to the
fubfequent eftablifhment at *Kufan*, or *Chufan*, in 1700, on which ifland he refided
fome time.

I conjecture alfo, that he was afterwards
fixed at *Pulo Condore*, and was the perfon
to whom we owe the account of the maf-
facre of the *Englifh*, by the *Macaffars*, at
that factory, in 1705, as related in the Modern Part of the *Univerfal Hiftory*, vol. x.
p. 154; edit. 1759. 8°.

Mr. CUNNINGHAM kept a journal of the
weather in both his voyages to *China*; and
during his refidence on the ifle of *Chufan*,
he appears to have been very active in collecting the productions of that place.   He
fent over to PLUKENET and PETIVER a
very large number of new plants, as is evident from an infpection of their writings.
In the " *Amaltheum*" of PLUKENET, his
name occurs in almoft every page.

From the ifland of *Afcenfion*, Mr. CUN-
NINGHAM tranfmitted to PETIVER an account of the plants, and fhells, he obferved
there.   He fent a journal of his voyage,
                                            and

and an account of the Iſle of *Chuſan*, which
was printed in the *Philoſophical Tranſactions,*
N° 280. vol. xxiii. p. 1201. It conveys
many intereſting particulars to the *Engliſh*
reader, relating to the inhabitants, their
fiſheries, agriculture, and arts. He corrects
ſeveral miſtakes of Father MARTINI, and
LE COMPTE; and is, I believe, the firſt
Engliſh writer, who gives an accurate hiſtory of the *Tea Tree:* which, although but
ſhort, is authenticated by the lateſt deſcription given us by THUNBERG, in the "*Flora Japonica.*"

Beſides this account of *Chuſan*, I find
the following papers, written by Mr. CUN-
NINGHAM, and printed in the *Philoſophical
Tranſactions.*

A Catalogue of Plants and Shells, collected on the Iſle of *Aſcenſion.* N° 255.
vol. xxi. p. 295.

Obſervations on the Weather at *Emuy*,
in *China*, in 1698, 1699; with the State
of the Barometer. N° 256. vol. xxi.
p. 323.

On the Declination of the Needle, and a
Thermometrical

Thermometrical Obfervation, near the Line.
N° 264. vol. xxii. p. 577.

A Meteorological Regifter of the Weather, in a Voyage to *China*, in 1700 ; and
a Regifter of the Weather at *Chufan*, in
China.   N° 292. p. 1639, and 1648.

BROWN.

Befides Mr. CUNNINGHAM, there were
two ingenious furgeons of the name of
BROWN, refident in the *Eaft Indies*, contemporary with PLUKENET, and PETIVER,
to whom thefe writers were under fingular
obligations, for the liberality and importance of their communications, both of vegetable and animal productions, from the
Eaft Indies.

Mr. *Samuel* BROWN was furgeon to the
Fort at *St. George*, now called *Madras*.
His correfpondence with Mr. PETIVER,
and his collections, have been noticed before.

The name of Mr. *Alexander* BROWN occurs in many parts of PLUKENET's works.

He

He difcovered feveral new plants, both in the *Eaſt Indies,* and at the *Cape of Good Hope* *.

G L E N.

Among the contemporaries of Mr. RAY, I am led, from private information, to mention *Andrew* GLEN, M. A. a divine, who, although his name does not occur in the writings of his time, was the friend and acquaintance of that eminent man ; having probably acquired a taſte for ſimilar purſuits, from a frequent intercourfe with him, at the feat of his illuſtrious friend, Mr. WILLUGHBY, near *Nottingham.* I have feen an *Herbarium,* collected by Mr. GLEN, which bears date in 1685, containing upwards of feven hundred indigenous, and two hundred exotic plants. Some of thefe

---

* Dr. PLUKENET denominated a new genus of *African* plants belonging to the *Pentandrous* claſs, *Eriocephalos Bruniades,* in honour of *Alexander* BROWN. LINNÆUS has perpetuated the genus ; but, agreeably to his rules, which do not admit of fuch terminations, has changed the term to BRUNIA, of which fome fpecies are known in the *Engliſh* gardens.

were

were collected in the tour on the continent.
He afterwards travelled into *Sweden* and
Italy; and refided fome time at *Turin*, where
he began another collection, which is dated
1692; but contained not more than two
hundred fpecimens. This *Herbarium*, all
circumftances confidered, was not mean
for the time in which it was made. Mr.
GLEN was afterwards rector of *Hatherne*,
in *Leiceflerfhire*, where he died at an advanced age.

C H A P. 32.

Sloane—*Memoirs of*—*Born in* Ireland—*His strong bias to natural history in his youth*—*Travels with Dr.* Tancred Robinson—*Favourite with* Sydenham—*Attends the Duke of* Albemarle *to* Jamaica—*Successful in his pursuit of objects in natural history*—*Secretary to the Royal Society*— *Zealous promoter of the dispensary*—Catalogus Plantarum Jamaicæ—Sloane's *liberal communication to* Ray—*Greatly augments his Museum by the accession of* Courten's.

S L O A N E.

AT the fame period of time, lived RAY, MORISON, PLUKENET, PE-TIVER, SLOANE, and SHERARD, under whofe countenance, and culture, the knowledge of nature received the moft rapid and fubftantial improvement, which it had ever experienced. In this period, fyftem had been revived and improved by MORISON, RAY, HERMAN, TOURNEFORT, and RI-VINUS. In this period alfo, RHEDE, RUM-PHIUS, PLUMIER, and SLOANE, publifhed

VOL. II.   F   thofe

thofe great works in exotic botany, which
have immortalized their names. This period was the clofe of the laft century;
which, as it has been called by the elegant
and learned author of the "Effay on the Ge-
" nius and Writings of Pope," " the Gol-
" den Age of Learning in *England*;" fo has
LINNÆUS named it, in his Allegorical Hiftory of the Rife and Progrefs of this Science, " The GOLDEN AGE OF BOTANY;"
and SLOANE was one of its brighteft ornaments.

Of the life of this great patron of natural
fcience, it would be fuperfluous in me to
attempt a detailed account; fince this tribute has been paid to his memory in the
" Eloge of the *French* Academy," in the
" *Biographia Britannica*," the " Biographical
" Dictionary," and other collections of that
kind, in daily ufe. Hence, I fhall, from
thefe publications, extract only the outlines
of his life, as they are connected with, and
tend to elucidate, his general character, his
acquirements in natural hiftory, and his botanical publications.

Sir *Hans* SLOANE was defcended from
parents,

parents, originally of *Scottish* extraction, and
was born at *Killileagh*, in the county of
Down, in *Ireland*, April 16, 1660. His
younger years were marked by a strong attachment to the works of nature. At six-
teen, his studies were interrupted by ill
health, in consequence of a spitting of blood,
which confined him for three years. On
his amendment, he studied the preliminary
branches of physic in *London*, for four
years; chemistry, under a pupil of the great
STAHL; his favourite science, at *Chelsea*
Garden, then but just established; and,
young as he was, contracted during that
time, an acquaintance with BOYLE and
RAY. Mr. SLOANE afterwards, in company with Mr. *Tancred* ROBINSON, and
another student, visited *France* for improvement. At *Paris*, he attended TOURNE-
FORT and DU VERNEY; and is supposed to
have taken his degrees in medicine at *Montpelier*; some say, at *Orange*. He returned
to *London* at the latter end of the year 1684,
and became the favourite of Dr. SYDEN-
HAM, who took him into his house, and zea-
lously promoted his interest. In November

F 2 1684,

1684, he was elected a fellow of the *Royal Society*; and, in April 1687, entered into the college of phyficians. Thefe early advancements in the profeffional line, are the ftrongeft prefumptions in favour of his fuperior knowledge, and promifing abilities. Yet thefe flattering profpects he relinquifhed, to gratify his ardour for natural knowledge.

On the 12th of September 1687, and in the 28th year of his age, he embarked for *Jamaica*, as phyfician to the Duke of *Albemarle*; and touched at *Madeira*, *Barbadoes*, *Nevis*, and *St. Kitt's*. The Duke dying on the 19th of December, foon after their arrival at *Jamaica*, Dr. SLOANE's ftay on the ifland did not exceed fifteen months. During this time, however, fuch was his application, that, in the language of his *French* eulogift, had he not converted, as it were, his minutes into hours, he could not have made thofe numerous acquifitions, which contributed fo largely to extend the knowledge of nature; while they laid the foundation of his own future fame and fortune.

Here I am led to obferve, that feveral circumftances

circumftances concurred refpecting the voyage of Dr. Sloane to *Jamaica,* which
rendered it peculiarly fuccefsful to natural
hiftory. He was the firft man of learning,
whom the love of fcience alone had led from
England, to that diftant part of the globe;
and, confequently, the field was wholly open
to him. He was already well acquainted
with the difcoveries of the age. He had an
enthufiafm for his object, and was at an age,
when both activity of body, and vivacity
of mind, concur to vanquifh difficulties.
Under this happy coincidence of circumftances, it is not ftrange that Dr. Sloane
returned home with a rich harveft. In fact,
befides a proportional number of fubjects
from the animal kingdom, he brought from
Jamaica, and the other iflands they touched
at, not fewer than 800 different fpecies of
plants. A number, very far beyond what
had been imported, by any individual into
England before.

Dr. Sloane returned from his voyage
on the 29th of May 1689; and, fixing in
London, foon became eminent. He was
chofen phyfician to Chrift's Hofpital, in

F 3 1694.

1694. In the preceding year, he had been elected secretary to the *Royal Society,* and had revived the publication of the *Philosophical Transactions,* which had been interrupted from the year 1687. This office he held till 1712, and was then succeeded by Dr. HALLEY.

In 1695, Dr. SLOANE married *Elizabeth,* daughter of Alderman *Langley,* of *London.* She died in 1724. She brought him, besides a son, and daughter, who died young, two other daughters, who survived, and were advantageously married; the eldest to *George Stanley,* Esq; of the county of *Hants;* and the younger to Lord *Cadogan.*

It was about this time that he became an active member of the college, in promoting the plan of a dispensary for the poor; which was at length carried into effect. The feuds excited on this occasion gave rise to the celebrated satire from Dr. GARTH.

It was not till the year 1696 that Dr. SLOANE published the *Prodromus* to his History of *Jamaica* Plants, under the title of " CATALOGUS PLANTARUM QUÆ IN
                                                INSULA

INSULA JAMAICA SPONTE PROVENIUNT,
*vel vulgo coluntur; cum earundem Synonymis
et Locis natalibus; adjeĉtis aliis quibuſdam
quæ in Inſulis Maderæ, Barbados, Nieves, et
Sanĉti Chriſtophori naſcuntur : ſeu Prodromi
Hiſtoriæ Naturalis Jamaicæ Pars Prima."*
8°. 1696. pp. 232. *Præter Indicem valde
copioſum Nominum et Synonymorum.*

This volume, intrinſically valuable as it
is, may yet be conſidered as only the nomenclature, or ſyſtematic index to his ſubſequent work. The arrangement of the
ſubjeĉt (and which was ſtriĉtly followed in
" The Hiſtory,") is nearly that of Mr. RAY;
vegetables being thrown into twenty-five
large natural claſſes, or families. Among
botaniſts of that time, generical charaĉters
had not attained any remarkable preciſion;
and SLOANE, like PLUKENET, was little
farther anxious, than to refer his new plants
to ſome genus already eſtabliſhed, without
a minute attention to the parts of fruĉtifi-
cation, farther than as they formed part of
the charaĉter drawn from habit: yet with
this defeĉt, the figures and deſcriptions of
SLOANE proved ſufficiently accurate to enable

able his fucceffors to refer almoft all his
fpecies, to the appropriate places in the fyftem of the prefent day.

By this neglect of conftructing *genera,*
SLOANE neverthelefs threw into the hands
of PLUMIER the grateful opportunity which
he embraced, of naming the plants of his
inveftigations after celebrated botanifts. In
juftice, however, to PLUMIER, it has been
before obferved, that he was not parfimo-
nious in the diftribution of thefe favours, to
the merit of *Englifhmen.*

It is worthy of obfervation, that among
thefe claffes, there are only two plants belonging to the *Umbelliferous* tribe, and but
one genus of the *Afperifoliæ,* namely, the
Heliotropium. The ferns, on the other
hand, are very numerous all over the *Weft
India* iflands. SLOANE has above one hundred fpecies ; and PLUMIER, a few years
afterwards, detected many more.

In this volume, however fmall in bulk,
yet vaft in labour, there is a circumftance
much to the credit of SLOANE, which muft
be obvious to every intelligent naturalift. It
is the care which the author has taken to
confult

confult every poffible refource, in order to
difcriminate his plants, and avoid an unneceffary multiplication of fpecies, by defcribing that as new, which was before known.
So numerous a fet of fynonyms had never
been inferted in any local catalogue; and
SLOANE greatly enhanced its value, by a
moft commendable addition; having, with
incredible labour, referred to every traveller
of note for all the vegetables renowned for
utility in medicine, arts, or œconomy. In
this inftance, it is much to be regretted
that fo praife-worthy an example has not
been more frequently imitated by fucceeding botanifts.

Before I difmifs this volume, I muft
mention an inftance of the liberality of its
author, in allowing Mr. RAY the free ufe
of his manufcripts of the *Jamaica* plants,
on the publication of the third volume of
his " Hiftory," in 1704. Accordingly, we
find all SLOANE's new plants, with the defcriptions at large, inferted in that work.
He alfo communicated a lift of *Englifh*
plants, which he had obferved fpontaneoufly
growing in *Jamaica*. This was printed in
the

he fecond edition of the *Synopfis*, and continued by DILLENIUS in the third.

Dr. SLOANE began early to form a mufeum, and it was, by the collections made
in his voyage, become confiderable; but the
æra of its celebrity was not till 1702, when
it received the augmentation of Mr. COUR-
TÈN's valuable ftores.   United by fimilar
tafte, Dr. SLOANE had formed, with this
gentleman, an early and ftrict friendfhip;
and a perpetual interchange of communications, and good offices, had fubfifted between them; of which Sir *Hans* himfelf bears
public teftimony in his writings.   It is not
poffible, at this time, to afcertain the extent of Mr. COURTÈN's collection; but it
is fufficiently certain that it was very ample:
the acquifition of it added new ardour and
diligence to our naturalift, in the augmentation of it.   He has himfelf exhibited a general ftatement of the contents of his mufeum, twenty-two years after its enlarge-
ment by the above-mentioned collection.
By this it appears, that the fubjects of natural hiftory alone, exclufive of two hundred
volumes of preferved plants, amounted to
more

more than 26,200 articles. They were afterwards augmented to upwards of 30,600;
as may be feen by " A General View of the
Contents," publifhed a year before his death.
—And here I cannot but obferve, that the
curious are under fingular obligation to the
author of the article COURTEN, in the fourth
volume of the new edition of the *Biographia Britannica,* who has, with great labour,
brought to light fo many interefting memoirs relating to Mr. COURTEN, and his
family. His truly laborious refearches have
done equal juftice to that gentleman, and to
Sir *Hans* SLOANE, by refcuing the hiftory
of their connexion from great mifreprefent-
ation. Hence we learn, that Dr. SLOANE,
as executor to his friend, fo far from ac-
quiring the acceffion of Mr. COURTEN's
mufeum *at a dear rate,* as had been reprefented in the preceding edition, obtained it
at a price greatly under its original, and real
value.

CHAP.

C H A P. 33.

Continuation of Sloane—*Publishes his voyage to*
Jamaica—*His discoveries excite emulation—*
Corresponds with Ray—*Elected member of the*
Royal Academy of Paris—*Chosen physician to*
Queen Ann—*Created a baronet—Physician ge-*
neral to the army—President of the College of
Physicians—and president of the Royal Society—
Retires to Chelsea *in* 1741—*His death and*
character—List of his numerous memoirs in the
Philosophical Transactions.

S L O A N E.

IN the year 1701, Dr. Sloane was incorporated doctor of physic at *Oxford* ;
and was associated member of several aca-
demies on the continent. In 1707, he
published the first volume of his history,
under the following title:

" A Voyage to the Islands *Madeira,*
" *Barbadoes, Nevis, St. Christopher's,* and
" *Jamaica* ; with the Natural History of
" the Herbs and Trees, four-footed Beasts,
" Fishes, Birds, Insects, Reptiles, &c. To
" which

" which is prefixed an Introduction, where-
" in is an Account of the Inhabitants, Air,
" Waters, Diseases, Trade, &c. of that
" Place; with some Relations concerning
" the neighbouring Continent and Islands
" of America." Vol. i. 1707. fol. pp.
254. tab. 156.

This volume is dedicated to the queen.
The introduction contains 154 pages, and
is replete with matter of an interesting nature, and evidently displays the great pains
the author took to collect materials for this
work. It comprehends a general account
of the discovery of the *West Indies*, and of
the island of *Jamaica* in particular: the geo-
graphy, the climate, and soil; food, manners, and economy of the various inhabitants: description of a tour the author made
to various parts of the country, and especially to the ruins of *Sevilla*, with an account
of the church built by *Peter Martyr*: a list
of more than fifty European vegetables,
principally of the culinary tribe, cultivated
in the gardens of *Jamaica*. He concludes
with an history of the diseases, and the detail of many cases which came under his

own

own obfervation. Then follows the journal of the voyage; with ample defcriptions
of the marine animals obferved in the courfe
of it: the plants obferved at *Madeira*, feveral of which are figured in the work itfelf:
obfervations of the like kind made at *Barbadoes*. The remainder of the volume contains a methodical arrangement, and defcription, of all the fubmarine, and herbaceous plants, natives of the ifland; amounting to 550 and upwards. Very few fynonyms are here introduced; but references,
in general, made to the copious collection
of them in his " Catalogue." To each
plant the author has fubjoined, from medical authors, and from travellers, the moft
ample account of their feveral ufes.

The *fecond volume* was not publifhed till
the year 1725, and was dedicated to the
king. The reafons of this delay are affigned
in the introduction, and were, principally,
the care, the arrangement, and defcription
of his mufeum. The acceffion of Mr.
COURTEN's collection has been remarked
above; and that of PETIVER, in 1718,
gave Sir *Hans* much employment. PETI-

ver had amaffed a greater quantity of the
productions of nature, than any man before him: but he had not preferved them
with a care equal to the zeal with which he
acquired them; and it demanded extraordinary diligence to recover them from the in-
jury they had fuftained. It is in the introduction to this volume that Sir *Hans* gives
a general inventory of his Library, and Mufeum, as it ftood in the year 1725, which
has been noticed before; and, by the comparifon of which with later eftimates, it appears how greatly he increafed it after that
time.

This fecond volume comprehends five
hundred pages, and completes the vegetable
part, and the animal kingdom. The new
plants are nearly all figured. The plates
are continued to the number of 274. The
laft forty belong to the animals, of which,
fome of all the claffes, the *Mammalia* ex-
cepted, are here exhibited.

To the curious botanift, it will be ob-
fervable, that out of 800 vegetables, defcribed in thefe volumes, above 100 are
Ferns; and that of the others, more than
250 fpecies are of the *arborefcent* kind.

Subfequent

Subfequent voyagers have eftablifhed it as a fact, that in the warmer and intertropical regions, this latter clafs conftitutes, in a general way, the third part of the vegetable productions of nature. Abundantly the reverfe of this takes place in temperate and cold climates.

In thefe volumes Sir *Hans* has introduced all his quotations at length from the books of travels mentioned in the " Catalogue," to illuftrate the various ufes of each vegetable. They exhibit a proof of the author's induftry, which, I conceive, it is difficult to parallel in any other work. The tables, which were principally engraved by *Vander Gucht*, although on a large fcale, yet having the difadvantage of being done from dried fpecimens, want thofe natural attitudes which grace more modern performances. In this volume, Sir *Hans* takes various occafions to defend himfelf from the ftrictures of PLUKENET; and, in his turn, criticifes that author, though in a language much lefs cenfurable than that of the *Mantiffa*.

The voyage of Dr. SLOANE was productive of much fubfequent benefit to fcience,

by

by exciting an emulation, both in *Britain*
and on the continent. Sir *Arthur* RAW-
DON, upon viewing his fplendid collection,
fent *James* HARLOW, a fkilful gardener,
to *Jamaica,* who returned with a fhip almoft laden with plants, in a vegetating
ftate; and with a great number of dried fpecimens. Of the latter, SLOANE had all
fuch as were new, before he publifhed his
firft volume. Many of the living plants
fucceeded in the garden of Sir *Arthur,* at
Moyra, in *Ireland;* and many were diftri-
buted into the garden of the Bifhop of *London,* at *Fulham;* Dr. UVEDALE's, at *Enfield;* the *Chelfea* Garden; and efpecially
into that of her Grace the *Duchefs* of BEAU-
FORT, at *Badminton,* in *Gloucefterfhire:*
the botanic gardens of *Amfterdam, Leyden,
Leipfic,* and *Upfal,* fhared thefe rarities.
TOURNEFORT fent Dr. GUNDELSCHEI-
MER, his affociate in his oriental journey,
into *England,* to view SLOANE's plants;
and this gave occafion to PLUMIER's expedition into the *Caribbee* Iflands.

Dr. SLOANE entered very early into correfpondence with Mr. RAY. His firft let-

ter bears date in 1684; and DERHAM's collection contains thirteen. Moft of them have
reference to botanical fubjects, while they
evidence the mutual friendfhip of the writers; a cirumftance indeed very pathetically
expreffed by Mr. RAY, in the laft letter he
ever wrote; which was but a few days before his death, being dated Jan. 7, 1704.

In 1708, he was elected a foreign member of the *Royal Academy* of *Sciences* at *Paris*; a diftinction of the higheft eftimation
in fcience; and the greater at that time, as
the *French* nation was at war with *England*, and the queen's confent was neceffary
to the acceptance of it. He was frequently
confulted by Queen *Ann*, who, in her laft
illnefs, was blooded by him. On the acceffion of *George* I. he was created a baronet; being the firft *Englifh* phyfician on
whom an hereditary title of honour had
been conferred. He was appointed Phyfician General to the Army, which office he
enjoyed till 1727, when he was made Phyfician to King *George* II. He gained the
confidence of Queen *Caroline*, and pre-
fcribed for the royal family till his death.

                                    In

In 1719, Sir *Hans* was elected Prefident of the College of Phyficians, which ftation he held fixteen years, and during that time he gave fignal proofs of his zeal for the interefts of that body.

On purchafing the manor of *Chelfea*, he gave the fee fimple of the *Botanical Garden* to the Company of Apothecaries, on conditions, which will more properly be fpecified in a fubfequent part of this work.

On the death of Sir *Ifaac* NEWTON, in 1727, Sir *Hans* SLOANE was advanced to the prefidency of the Royal Society of *London*, the intereft of which no man had ever more uniformly promoted. He made the Society a prefent of 100 guineas, and a buft of the founder, King *Charles* II. Thus, in the zenith of profperity, he prefided, at the fame time, over the two moft illuftrious fcientific bodies in the kingdom : and, while he difcharged the refpective duties of each ftation, with credit and honour, he alfo enjoyed the moft extenfive and dignified employment as a phyfician. He occupied thefe important ftations from the year 1719 to 1733, when he refigned the prefidency of the College of Phyficians ; and, in 1740, at the age of

fourfcore,

fourfcore, that of the *Royal Society*; having formed the refolution of withdrawing from the buftle of life into retirement at *Chelfea*. In 1741, he removed his Library, and Mufeum; and, on the 12th of May, fixed at his new manfion, where, occafionally vifited by his friends, and by all men of diftinction from abroad, he paffed in ferenity; and in the conftant exercife of benevolence, the laft twelve years of his life, which terminated Jan. 11, 1752, O. S.

Sir *Hans* SLOANE was tall, and well made in his perfon; eafy, polite, and engaging in his manners; fprightly in his converfation; and obliging to all. To foreigners he was extremely courteous, and ready to fhew and explain his curiofities to all who gave him timely notice of their vifit. He kept an open table once a week for his learned friends, particularly thofe of the Royal Society. In the aggregation of his vaft collection of books, he is faid to have fent his duplicates, either to the Royal College of Phyficians, or to the Bodleian Library.

He was governor of almoft every hofpi-

tal

tal in *London* ; and to each, after having given an hundred pounds in his life-time, he left a more confiderable legacy at his death. He was ever a benefactor to the poor, who felt the confequences of his death feverely. He was zealous in promoting the eftablifhment of the colony of *Georgia*, in 1732 ; and formed, himfelf, the plan for bringing up the children in the Foundling Hofpital, in 1739.

In the exercife of his function as a phy- fician, he is faid to have been remarkable for the certainty of his prognoftics ; and the hand of the anatomift verified, in a fignal manner, the truth of his predictions, relat- ing to the feat of difeafes. By his practice, he not only confirmed the efficacy of the *Peruvian Bark* in intermittents, but extend- ed its ufe in fevers of other denominations, in nervous diforders, and in gangrenes and hemorrhages. The fanction he gave to in- oculation, by performing that operation on fome of the royal family, encouraged, and much accelerated its progrefs throughout the kingdom. His ointment for the *Leu-*

G 3 *coma*

coma has not yet loft its credit with many reputable names in phyfic.

Poffeffed of affluence, entirely the reward of his own merit, Sir *Hans* SLOANE enjoyed, through a long life, every fatisfaction that fcience could beftow; and, in the *Britifh Mufeum*, has not only erected the nobleft monument to his own fame, but a temple alfo to the culture of learning and of fcience, which will tranfmit his donation to future ages, as a fignal inftance of the munificence of a private individual.

That the lift of Sir *Hans* SLOANE's writings may be complete, I fhall, in conformity with my general plan in other inftances, recite thofe papers which were printed in the *Philofophical Tranfactions*. Many of thefe convey fuch interefting literary information, and abound with fuch facts and obfervations, as will long give them a value, with all who remember the eminent abilities and fervices of the author, and have a relifh for the like refearches.

The firft of Dr. SLOANE's papers in the *Philofophical Tranfactions*, is, a Defcription
of

of the *Jamaica* Pepper Tree *(Myrtus Pi-menta* Lin.)*;* with an account of curing the Berries; and of the Oil diftilled from them, commonly fold for *Carpobalfamum.* N° 192. Vol. xvii. p. 462; accompanied with a figure.

A Defcription of the Wild Cinnamon Tree, falfely called *Cortex Winteranus (Win-terania Canella* Lin.) very early celebrated, by the firft writers on the Weft Indies, as a medicine for the Scurvy. N° 192. Vol. xvii. p. 465.

A Defcription of the Silver Pine *(Protea Argentea* Lin.)*;* and of another Coniferous Tree; both from the *Cape of Good Hope.* N° 198. Vol. xvii. p. 664.

Proofs of the Poifonous Effects of Dog's Mercury *(Mercurialis Cynocrambe,* Lin.) N° 203 Vol. xvii. p. 876.

An Account of the true *Cortex Wintera-nus (Drimys Winteri* Lin. Sup. p. 269.) from the *Straights of Magellan.* Dr. SLOANE obferves, that the fenfible quali-ties of this bark are fo fimilar to thofe of the *Canella,* that he judges they may fafely

be

be confidered as *fuccedanea* to each other.
N° 204. Vol. xvii. p. 922 ; with a figure
of a fmall branch.

An Account of the Coffee Shrub, from
the firft fpecimen brought over from *Ara-
bia Fœlix* by Mr. CLIVE; with a figure, and
an account of the culture of the fhrub.
N° 208. Vol. xviii. p. 65.

An Account of the Bird called the *Con-
dor* of *Peru*, from the relation of Captain
Strong, who had met with one on the coaft
of *Chili*, which meafured 16 feet from tip
to tip of the wings. This is one of the
earlieft accounts of this bird, after that of
De Laet; concerning the ftrength and rapa-
city of which, voyagers have related incredible ftories. LINNÆUS names it *Vultur
Gryphus*. N° 208. Vol. xviii. p. 61.

An Account of an Earthquake which
happened during Dr. SLOANE's ftay on the
Ifland of *Jamaica*, Feb. 19, 1687-8. With
a Note concerning the great Earthquake of
June 7, 1692, which deftroyed *Port Royal*.
N° 209. Vol. xviii. p. 80, 81.

Remarks on a vulgar Opinion that the
                                    fwallowing

fwallowing of Stones affifts Digeftion; occafioned by the cafe of a man who had
fwallowed 200 pebbles. N° 253. Vol.
xix. p. 192.

Obfervations on four Kinds of *American*
Fruits, thrown on the fhores of the North-
weft parts of *Scotland*. Three of thefe were
known by the author to be natives of *Ja-
maica.*—1. The Cocoons, or the Pods of
the *Phafeolus Maximus,* &c. *Hift. Jam.* i.
p. 178. *(Mimofa Scandens* Lin.)—2. The
Horfe Eye Bean; *Phafeolus Brafilienfis,* &c.
Hift. Jam. i. p. 178. *(Dolichos Pruriens*
Lin.)—3. The Afh-coloured Nickar Tree;
Lobus Echinatus, &c. *Hift. Jam.* ii. p. 40.
(Guilandina Bouduc Lin.)—4. The *Fructus
Orbicularis fulcis nervifque diftinctus,* C. B.
pin. 405. *b.* iv. N° 222. Vol. xix. p. 298.
Thefe, and feveral other kinds, which are
alfo found plentifully on the coaft of *Nor-
way,* are thought by SLOANE to have been
brought by currents, through the *Gulph of
Florida,* into the *North American* ocean.
The appearance of thefe fruits on the Nor-
thern fhores of *Europe,* had been alledged
by

by some as proofs of the existence of a North-east passage.

An Account of the Fossil Tongue of a *Pastinaca Marina (Raia Pastinaca* Lin.) dug up in *Maryland*; with a Comparison of it with the recent Tongues of the *Thornback*; illustrated with many figures. N° 232. Vol. xix. p. 674.

Remarks on *Dampier*'s Medicine for the Bite of a Mad Dog, specifying, that it was not a *Jew's Ear Fungus*, but the *Lichen Cinereus Terrestris* of RAY. N° 237. Vol. xx. p. 52.

Notes on a Paper, written to recommend the *Ipecacuanha*, as an infallible Remedy in Dysenteries. Dr. SLOANE recommends it, but speaks in a more moderate stile concerning its efficacy. He shews that it was first mentioned by an anonymous *Portuguese*, under the name of *Ipecaya*, or *Pigaya*. See *Purchas*'s *Pilgr. vol.* iv. N° 238. Vol. xx. p. 78.

An Account of the Contents of a *China* Cabinet, containing Instruments and Natural Curiosities; presented to the Royal Society

ciety by Mr. BUCKLEY, Surgeon at *Fort
St. George.* N° 246—250. Vol. xx. and
xxi.

An Account of the *Tartarian* Lamb,
Agnus Scythicus, or *Barometz,* heretofore
impofed on the credulous as a kind of Zoo-
phyte, or vegetating Animal. On examination, it proves to be the lower part of the
ftipes, or root, protruding above ground, of
a large fpecies of Fern, judged by fome to
be the *Polypodium Aureum,* fafhioned into
the rude fhape of the animal. It is naturally cloathed with a *lanugo,* or down, of
a yellowifh chefnut colour, called *Poco-
fempie* in *China* and the Eaft, celebrated as
an aftringent, both internally and externally
ufed; with a figure of the pretended animal. N° 247. Vol. xx. p. 461.

An Account of the *Nux Pepita,* or *St.
Ignatius'* Bean *(Ignatia Amara* Lin. Sup.
149); a fimple in common ufe in the *Philippine* Iflands, as a tonic medicine. N° 249.
Vol. xxi. p. 44.

An Account of fome Seeds, ufed on the
coaft of *Malabar* and *Coromandel,* for clarify-

ing

ing Water. N° 249. Vol. xxi. p. 44. There can be little doubt that this effect arifes from the mucilaginous quality.

A Defcription, with the Figure, of a Miffeltoe, or *Epidendrum*, called Wild Pine, in *Jamaica*, *(Tillandfia Utriculata* Lin.) the leaves of which form a refervoir for water, of great ufe in dry feafons. With Obfervations on the Oeconomy of feveral other Vegetables in the Propagation of the Species. N° 251. Vol. xxi. p. 113.

Hints for the Improvement of Gardening, to be drawn from due attention to the nature of the foil and climate, &c. of the plants. N° 251. Vol. xxi. p. 119.

The Method ufed by the *Chinefe* to make Gold Thread, by gilding paper on one fide with leaf gold, cutting it into long pieces, and weaving it into their filks, which makes them, with little or no coft, look very rich and fine. N° 251. Vol. xxi. p. 71.

A Limeftone Marble, difcovered in *Wales* by Dr. LHWYD, determined by Dr. SLOANE to be the *Aftroites* of WORMIUS. N° 252. Vol. xxi. p. 188. (Since called *Corallia Aftroita.)*

Aftroitæ.) It is native in the feas of *Ja-maica.*

The Cafe of a Dropfy of the *Ovarium.* N° 252. Vol. xxi. p. 150.

The Mifchief of fwallowing Plumb Stones exemplified, in the cafe of a man, in whofe bowels was found a plumb ftone inclofed in the center of a *calculus ægagropila,* which meafured fix inches in circumference, and weighed one ounce and an half. N° 282. Vol. xxiii. p. 1283.

An Account of the Moffes, or Turf Bogs, in the North of *Ireland.* N° 330. Vol. xxvii. p. 296.

Remarks on Mr. S. GRAY's Account of the Foffils of *Reculver Cliff;* tending to prove that the wood found there is that of the Oak. N° 368. Vol. xxii. p. 762.

An Account of a Pair of very extraordinary large Horns, found in a cellar at *Wapping;* with figures. Dr. HOOK fufpected they were the horns of an animal, defcribed by NIEU-HOFF, under the name of *Sukotyro,* as it is called by the *Chinefe.* Sir *Hans* conjectures, they might belong to the *Taurus carnivorus* of *Agatharchides;* of which he traces the
hiftory

hiftory through the writings of the antients;
but thinks it very uncertain whether this
is the fame animal with the *Sukotyro*. N°
397. Vol. xxxiv. p. 222.

An Account of fuch Specimens of Ele-
phants Teeth, and Bones, as are repofited
in the Mufeum of Sir *Hans* SLOANE; with
figures. N° 403. Vol. xxxv. p. 457. This
was introductory to

Remarks on divers Accounts of Teeth,
and Bones, found under ground. Ib. N°
404. p. 497. In which the curious reader
meets with much information.

An Account of the Symptoms arifing
from eating the Seeds of *Henbane*; with
Remarks. N° 429. Vol. xxxviii. p. 99.

Conjectures on the fafcinating Power attributed to the *Rattle-fnake*. N° 433. Vol.
xxxviii. p. 321.

Anfwer to the *Marquis of* CAUMONT's
Letter, relating to an extraordinary *Calculus*
of the Bladder. N° 450. Vol. xl. p. 374.
The ftone is figured in the *Tranfactions*. In
fhape, it refembled, in fome meafure, a globular *pyrites* befet with long, blunt, and
ramified points. N° 450. Vol. xl. p. 374.

                                  Anfwer

Anſwer to Mr. Powel, concerning the Caſe of a Gentlewoman, who voided with her Urine, hairy, cruſtaceous Subſtances ; informing him of ſimilar Caſes, and hinting a Method of Cure. N° 460. Vol. xli. p. 703.

The Deſcription, with a Figure, of a curious Sea Plant ; *Frutex Marinus flabelli- formis cortice verrucoſo obductus.* Doodii. Raii Syn. ed. 3. p. 32. *(Gorgonia Verru- coſa* Lin.) N° 478. Vol. xliv. p. 51. Small ſpecimens have been found on the ſhores of *Cornwall*; but it has elſewhere grown to the height of four feet.

Accounts of the pretended Serpent Stone, called *Pietra de Cobra de Cabelos*; and of the *Pietra de Mombazzo*, or the Rhinoceros Be- zoar : with the figure of a Rhinoceros with a double Horn. N° 492. Vol. xlvi. p. 118.

An Account of Inoculation, by Sir *Hans* Sloane, Bart. given to Mr. Ranby to be publiſhed anno 1736. Vol. xlix. p. 516. Curious as a record of the introduction of Inoculation into *England*; and valuable, as a proof, not only of the candour, and open- neſs of the author, but, as exhibiting a ſim-
plicity

plicity in the management, which it would have been happy for fociety, had it been univerfally adopted by fucceeding practitioners *.

* The name of SLOANE was given by PLUMIER to an arborefcent plant of the *Polyandrous* clafs, firft defcribed by MARCGRAAVE. It is fo nearly allied to the Chefnut tree, that MILLER, in his Dictionary, refers it to that genus. LINNÆUS, however, on the credit of LOEFLING, preferves PLUMIER's appellation, *Sloanea*; and has added another fpecies from CATESBY's *Carolina* Plants.

CHAP.

C H A P. 34.

Royal Society—*Its influence in promoting natural hiftory.*

Chelfea Garden — *Herborizations* — *Endowed by Sir Hans* SLOANE—*Highly advantageous to botany—Demonftrators*—Rand—Miller.

Celebrated gardens after the Revolution — Hampton Court —Badminton—Duke of Argyle's —Uvedale's.

Bifhop Compton — *brief account of — a patron of botany—Himfelf a real botanift—His fine garden at* Fulham—*Ufeful to* Ray, Plukenet, *and* Petiver.

Doody — *Not an author* — Cryptogamia *clafs greatly enlarged by him.*

R O Y A L S O C I E T Y.

AMONG thofe public inftitutions, which in a fingular manner invigorated, in this period, the fpirit of natural hiftory in *Englana,* the ROYAL SOCIETY of *London* claims the moft diftinguifhed notice. In its defign, as in its progrefs, it was the foftering parent, and guardian, of

VOL. II.   H   natural

natural knowledge. The collection of a
museum of natural curiosities, was one of
the objects in view; and such was the re-
spectability of the society, both as a body,
and in its individuals, that, through its
means, the whole nation may be said to
have amply contributed to its emolument.
All new objects of curiosity; all new books,
at home and abroad ; new discoveries
throughout all nature, incessantly offered
themselves; and thus, not only tended to remove the prejudices that too strongly prevailed against the studies of nature in that
age, but, in the event, excited a passion in
the public, which was so successfully exerted in improving, not natural history alone,
but real and useful science at large, that it
will not be considered as an exaggerated
encomium to assert, that more effectual
advancement was made by the influence
of this illustrious body in one century,
than had before taken place from the ear-
liest ages. Botany shared these benefits;
and the early volumes of the *Transactions*
abound in records of newly-discovered vegetables, and of newly-discovered proper-

ties

ties in that part of the creation. Experiments of various intention were inftituted by this learned body. Under their aufpices, the anatomy, and philofophy of plants, were illuftrated by GREW, and improved by HALES. Such memoirs in the *Philofophical Tranfactions* as more directly relate to my object, or were written by thofe whom I commemorate, have been already, or will be, briefly noticed in the courfe of thefe pages.

CHELSEA GARDEN.

I proceed further to obferve, that, among the affiftances which the fcience has received from public bodies of men, muft be mentioned alfo that which fprung from the Phyfic Garden, founded by the Company of Apothecaries at *Chelfea*; an inftitution which reflected the higheft honour on that refpectable fociety. This took place in the year 1673, although the infcription over the entrance imports, that the Garden was not effectually conftructed till the year 1686.

From the time of JOHNSON, the editor of GERARD, a cuftom had prevailed among the *London* Apothecaries to form a fociety

H 2   each

each fummer, and make excurfions to inveftigate plants. The *Itinera,* publifhed
by JOHNSON, may be confidered as the fruit
of fuch expeditions in his day. But they
had been difcontinued, as I apprehend, for
many years. After the foundation of the
Garden, this laudable practice was revived,
and the affociations fixed to ftated periods,
and put under regulations; the herborizings
being now diftinguifhed into private and
general. The firft, intended to excite a
tafte for botany among the apprentices of
the Company, begin on the fecond Tuefday
of April, and are held monthly, on the
fame day, till September inclufively, in fome
of the villages in the immediate neighbourhood of *London*. At the end of the feafon,
the premium of a copy of RAY's *Synopfis,*
(fince changed for Mr. HUDSON's *Flora
Anglica)* is prefented to the young man
who has been the moft fuccefsful in difcovering and inveftigating the greateft number of plants. The general herborization
is annual only, in July; when the Demonftrator, and others of the Court of Affift-
ants, belonging to the Company, make an
excurfion to a confiderable diftance from
the

the city, collect the ſcarce plants, and dine together near *London*; on which occaſion they are frequently accompanied by other gentlemen, who are fond of the ſame pur-ſuits.

This inſtitution at *Chelſea* was rendered more ſtable, and received permanency, from the liberality of Sir *Hans* SLOANE; who, ſoon after his purchaſe of the manor, in 1721, gave the freehold of the ground, near four acres, to the Company, on condition that the demonſtrator ſhould, in the name of the Company, deliver annually to the *Royal Society* fifty new plants, till the num-ber ſhould amount to 2000, all ſpecifically different from each other; the liſt of which was publiſhed yearly, in the *Philoſophical Tranſactions*. The firſt was printed in the year 1722, and the catalogues have been continued till the year 1773, at which time the number 2550 was completed. Theſe ſpecimens are duly preſerved in the archives of the Society, for the inſpection of the cu-rious.

Under excellent ſuperintendants, *Chelſea Garden* has flouriſhed; having been excelled, perhaps, by no public inſtitution of the

kind

kind in *Europe*, for the number of curious exotics it contains. Of this, the Dictionary, and Figures of the late Mr. MILLER, afford fufficient proofs. The advantages, indeed of this inftitution are obvious; and even Sir *Hans* SLOANE himfelf acknowledged his obligations to it in the early part of his life.

In juftice to the memory of thofe who have eminently filled the place of lecturers, and demonftrators, in the *Chelfea Garden*, I recite their names, from the time of Sir *Hans* SLOANE's donation; not being able to afcend above that period.

Ifaac RAND, apothecary, F.R.S.	} 1722—1739
Jofeph MILLER, apothecary,	1740—1746
John WILMER, M.D.	1747—1764
William HUDSON, F.R.S.	1765—1769
Stanefby ALCHORNE,	1770—1772
William CURTIS,	1773—

Mr. RAND publifhed, in 1730, in 8°. *Index Plantarum Officinalium Horti Chelfejani.* The lift contains 518 plants of the *Materia Medica*; and fpecifies the part of each ufed in phyfic. The fame author publifhed

publifhed *Horti Chelsejani Index Compendia-*
rius. 1739. 8°*.

Joseph MILLER publifhed, " *Botanicum*
" *Officinale;* or, A Compendious Herbal:
" giving an Account of all fuch Plants as
" are now ufed in the Practice of Phyfic ;
" with their Defcriptions and Virtues." 8°.
1722. pp. 466. The plants are alphabeti-
cally arranged according to the officinal
names. The Summary of the Virtues is, in
moft inftances, a tranflation from the *Phar-*
macologia of DALE.

Except in the *Orthotonia,* fubjoined to
Shipton's edition of *Pharmacopœia Bateana,*
the *Botanicum Officinale* of MILLER is the
only book on the fubject, not of very modern date, in which the ftudent is affifted
in the accentuation of the *Latin* names of
plants; but, from the confined nature of the
plan in thefe works, the benefit is not extenfive.

Among the favourable circumftances
which contributed to diftinguifh, or, I might

---

* *Houfton* gave the name of RAND to a Weft India
fhrub of the *Pentandrous* clafs, figured by *Sloane:* and it
is retained by LINNÆUS.

fay,

fay, helped to form, the *Golden Age of Botany*, before alluded to, was that growing tafte for the cultivation of exotics, which fprung up among the great and opulent, after the happy return of internal peace by the Reftoration. *Archibald Duke of* ARGYLE is faid to have been one of the firft, who was confpicuous for the introduction of Foreign Trees, and Shrubs. Mr. EVELYN, both by his writings, and example, encouraged the fame tafte. He cultivated a garden and plantations at *Saye's-Court*, near *Deptford*; and his *Sylva* remains a monument of his learning, and patriotic intentions, which cannot foon be fuperfeded.

After the Revolution, the Royal Gardens at *Hampton-Court* were rich in fine plants, and that at *St. James's* of no inconfiderable note, if we may guefs by the many new plants PLUKENET received from it. The *Duchefs of* BEAUFORT had a garden richly ftored at *Badminton*, in *Gloucefterfhire*. Dr. *Henry* COMPTON, Bifhop of *London*, another at *Fulham*; and many private gentlemen vied with each other, in thefe elegant and ufeful amufements. The gardens of
Dr.

Dr. UVEDALE, of *Enfield*; of Mr. DU BOIS, an *Eaſt India* merchant; of Mr. COURTEN, and others, were of the firſt notice; and afforded much aſſiſtance to the labours of RAY, PLUKENET, and PETIVER. The growing commerce of the nation, the more frequent intercourſe with *Holland*, where immenſe collections from the *Dutch* colonies had been made, rendered theſe gratifications more eaſily attainable than before; and, from all theſe happy coincidences, ſcience in general reaped great benefit.

BISHOP COMPTON.

Among thoſe I have juſt enumerated, Dr. *Henry* COMPTON, Biſhop of *London*, claims peculiar notice; ſince we learn from Mr. RAY, and PLUKENET, that he joined to his taſte for gardening, a real and ſcientific knowledge of plants; an attainment not uſual among the great, in thoſe days.

This eminent prelate, ſo diſtinguiſhed for his zeal in the cauſe of Proteſtantiſm, and for the active part he took in promoting the Revolution, was born in the year 1632. He was entered a nobleman of

Queen's

Queen's College, *Oxford*, in 1649, where he continued about three years; and afterwards travelled abroad. After the Reftoration, he entered into the army; but very foon quitted it for the church. In the year 1666, he became Rector of *Cottenham*, in *Cambridgefhire*; and, after poffeffing various emoluments in the church, was made Bifhop of *Oxford*, in the year 1674; and the year after, tranflated to the See of *London*, which he held to the time of his death, in the year 1713, at the age of 81.

In his retirement at *Fulham*, Bifhop COMPTON amply gratified himfelf in his favourite amufement. The circumftances of the times, above mentioned, aided by his own zeal, and a refidence of thirty-eight years on the fame See, enabled him, finally, to collect a greater variety of Green-houfe rarities, and to plant a greater variety of hardy Exotic Trees, and Shrubs, than had been feen in any garden before in *England*.

This repofitory was ever open to the infpection of the curious and fcientific; and we find RAY, PETIVER, and PLUKENET, in numerous inftances, acknowledging the
affiftances

affiftances they received from the free communication of rare and new plants out of
the garden at *Fulham*. Many of PLUKE-
NET's figures were engraved from fpecimens out of the Bifhop's garden; and fome
from a book of drawings in his poffeffion,
quoted under the name of *Codex Comptoni-
enfis*.

In the fecond volume of RAY's Hiftory
of Plants, p. 1798, we find a Catalogue of
fome new fpecies of Trees and Plants, obferved by the author in this garden. Thefe
were principally of *North American* growth.
The reader who is defirous of feeing a more
ample account of the garden at *Fulham*, is
referred to a relation of the ftate in which
it was found in the year 1751, written by
the late Sir *William* WATSON, and printed
in the 47th volume of the *Philofophical
Tranfaƈtions*.

DOODY.

If to any man in his day, not profeffedly
an author on the fubjeƈt, extraordinary
praife is due, for difcoveries in the indigenous botany, it muft belong to Mr. *Samuel*
DOODY,

DOODY, the contemporary and friend of
RAY, PLUKENET, and SLOANE, who all
bear teftimony to his merit. I regret the
want of more information relating to this
affiduous man; of whom I can only fay,
that he was born in *Staffordfhire*, was an
apothecary in *London*, and a fellow of the
Royal Society. He was chofen fuperinten-
dant, and demonftrator of the Garden at
Chelfea; an office he held for fome years before his death, which took place in 1706.

As Mr. DOODY lived in *London*, and
there is room to believe he was in very confiderable bufinefs, his excurfions could not
ordinarily extend far from that city. In its
neighbourhood, his diligence was beyond
any other example. He ftruck out a new
path in botany, by leading to the ftudy of
that tribe, which comprehended the imperfect plants, now called the *Cryptogamia*
clafs. In this branch he made the moft numerous difcoveries of any man in that age,
and in the knowledge of it ftood clearly unrivalled. The early editions of RAY's *Synopfis* were much amplified by his labours; and
he is reprefented by Mr. RAY, as a man

of

of uncommon fagacity in difcovering and difcriminating plants in general. The learned fucceffor of *Tournefort*, M. Jussieu, fpeaks of him as *inter Pharmacopæos Londinenfes fui temporis Coryphæus.* In truth, he was the Dillenius of his time.

There is a long lift of rare plants, many of them new, and firft difcovered by Mr. Doody, publifhed in the fecond edition of Ray's *Synopfis,* accompanied with obfervations on other fpecies. I alfo find, " The " Cafe of a Dropfy of the Breaft," written by him, and printed in the *Philofophical Tranfactions* in 1697. N° 224. Vol. xx. p. 77.

C H A P.

C H A P. 35.

L L H W Y D.

*E*DWARD Llhwyd was one of the
moft learned and celebrated antiquaries
of the laft century, and a fkilful naturalift.
According to Mr. *Wood,* he was the fon of
Edward Llhwyd, or *Lloyd,* of *Kidwell,* in
Carmarthenfhire ; but, as *Boyer* relates, of
Charles Llhwyd, of *Lanvordia,* in *Salop,* Efq.
He was born in 1670, and became a ftu-
dent of *Jefus College* in 1687, of which
Dr. Plot was a member, and under whom
Mr. Llhwyd was in a great meafure edu-
cated.

cated. On the refignation of Dr. PLOT, he fucceeded him as keeper of the *Afhmolæan* Mufeum, about 1690. He travelled repeatedly all over *Wales*; went through the North of *Scotland*; into *Ireland,* in which kingdom he feems to have made fome confiderable abode. He fpent fome time in *Cornwall,* and in *Britany* in *France,* in fearch of antiquities, and materials to carry on the extenfive defigns he had formed.

In all thefe journies he was conftantly attentive to the objects of natural hiftory, foffils, plants, and any remarkable phænomena of nature. Such of his remarks as are publifhed in the *Philofophical Tranfactions,* are full of curious and new information on thefe fubjects. His knowledge of the plants of his own country, and particularly thofe of *Wales,* juftly entitles him to remembrance in thefe pages, although he was not profeffedly an author on the fubject of them.

LLHWYD firft brought to light feveral of the rare plants of *Wales*; many of which, till of late years, were confidered as natives of no other part of Britain. He difcovered
feveral

feveral very fcarce fpecies in *Cornwall* : all
thefe he communicated to Mr. RAY, who
inferted them in the editions of his *Synopfis*,
with due acknowledgments. After having
made very large collections for the antiquities
of *Wales*, and formed great defigns in literature, he died before he could digeft them
into order for publication, in July 1709.

Exclufive of large communications, made
to Bifhop GIBSON's edition of *Camden*, on
the antiquities of *Wales*, he formed, from
the fruit of his own inveftigations, the *Lift
of Plants* fubjoined to the account of that
country.

He publifhed, " LITHOPHYLACII BRI-
TANNICI ICHNOGRAPHIA." 1699. 8°.
cum tab. 25. This work, which is a methodical catalogue of the figured foffils of
the *Afhmolean* Mufeum, confifting of 1766
articles, was printed at the expence of Sir
Ifaac NEWTON, Sir *Hans* SLOANE, and a
few others of his learned friends. As only
120 copies were printed, a new edition of it
was publifhed in 1760 by Mr. *Huddesford*;
to which were annexed feveral letters from
Mr. LLHWYD to his learned friends, on the

fubjeÆt of Foffils; his *Prælectio de Stellis Marinis Oceani Britannici, et Afteriarum, Entrochorum, et Encrinorum Origine.*

In Mr. RAY's correfpondence, we meet with thirteen letters, written by LLHWYD; all, except one, bearing date in 1690, and the four fubfequent years. They run chiefly on the fubjeÆt of Foffils. In OÆtober 1692, he informs Mr. RAY, that he was employed in drawing up a Catalogue of Mr. ASH-MOLE's *Legacy* of Books, Medals, and Pic-tures. In the lift of his printed works, we find *Catalogus Librorum Manufcriptorum in Mufeo Afhmoleano,* in 10 fheets, folio, without date. In 1707, he publifhed "AR-CHÆOLOGIA BRITANNICA," fol. Vol. I. *Gloffography.* A work by which he will long be remembered among the lovers of antiquity.

From private information, I have learned that Mr. LLHWYD left a very confiderable library, a large colleÆtion of manufcripts and fpecimens; that in the year 1728, thefe were all in the cuftody of Dr. FOWLKES, of *Lhanher,* who died foon after that time. They were undifpofed of ten years afterwards; but the printed books, which were of great worth, had been valued by a gentleman and two bookfellers, and the refufal of them offered to the Duke of *Bedford*. My intelligence reached no farther.

I conclude this account with a catalogue of Mr. LLHWYD's papers, publifhed in the *Philofophical Tranfactions*.

An Account of the *Lapis Amianthus*, or *Linum Foffile Afbeftinum*, found in the Northern part of *Anglefey*; with Mr. LLHWYD's Method of making it into Paper. N° 166. Vol. xiv. p. 223.

A Defcription, in *Latin*, of feveral regularly-figured Foffils; with Figures. Thefe were the *Siliquaftra, Bufonites, Gloffopetræ*, &c. N° 200. Vol. xvii. p. 746.

On the Swarms of Locufts which appeared in *Wales* in the year 1693; and an Extract from a Manufcript Hiftory of *Pembrokefhire*, relating to an immenfe number of Caterpillars, which confumed the produce of 200 acres of grafs in three weeks time, in the year 1601. N° 208. Vol. xviii. p. 45.

A Relation of a fiery Exhalation, or

Damp, at *Harleck*, in *Merionethshire*, in 1693 and 1694, which set fire to several stacks of hay, and proved fatal to many cattle. Mr. LLHWYD observes, that the founding of drums and horns, &c. repelled it from houses and stacks of hay, and th... at length, by this means, many were preserved. N° 213. Vol. xviii. p. 223.

Of an extraordinary Shower of Hail at *Pontipool*, in June 1697. N° 229. Vol. xix. p. 579.

Some Observations on Languages; in which Mr. LLHWYD assents to Mr. PEZRON's opinion, that the *Greek*, *Roman*, and *Celtic*, have one common origin. N° 243. p. 280.

Roman, *French*, and *Irish* Inscriptions; and Antiquities in *Scotland* and *Ireland*, with seven Figures. N° 269. Vol. xxii. p. 790.

On the Difference between the Fossils of *Essex*, and those of *Wales*, and *Ireland*; in the former *calcareous*, and in the latter *crystalline*. N° 291. Vol. xxiv. p. 1566.

On some strange Birds observed in *Wales*. N° 334. p. 464.

I 2 Observations

Obſervations made on Natural Hiſtory
in *Wales*: on Marcaſites: Quadrupeds:
Fiſh: and on *Welch* Manuſcripts. N° 334.
Vol. xxvii. p. 462.

On the Foſſils and Iron Mines of *Breck-
nockſhire*. p. 467.

In *Ireland*. A Stonehenge, near *Drogheda*:
Giants Cauſeway: Baſaltes on the Top of
Cader Idris: Antiquities, &c. N° 335.
Vol. xxvii. p. 503.

Antiquities and ſcarce Plants in *Ireland*.
N° 336. ib. p. 524.

Natural Curioſities in *Cornwall*. N° 336.
Vol. xxvii. p. 527.

Foſſils and Antiquities in *Wales*: *Welch*
Coins: Druids Beads: ſcarce Plants. With
Figures. N° 337. Vol. 28. p. 93.

Obſervations in Natural Hiſtory; and on
Antiquities in *Scotland*. N° 337. Vol.
xxviii. p. 97.

Deſcription and Figure of a remarkable
Sea Plant *(Tubularia indiviſa* Lin.) N° 337.
Vol. xxviii. p. 71.

LAWSON.

Among thoſe who diſtinguiſhed themfelves

ſelves for their knowledge in botany, without publiſhing profeſſedly on this ſubject,
Thomas LAWSON, by whoſe diſcoveries the
Engliſh Flora was enlarged, muſt not be
forgotten.

Of Mr. LAWSON I can only collect,
that he lived at *Great Strickland,* in *Weſtmoreland,* at the time he tranſmitted to Mr.
RAY " A Catalogue of the Rare Plants of
" the North of *England* ;" which is printed
in the " Philoſophical Letters," p. 213. This
liſt clearly evinces, that the author muſt
have paid great attention to his ſubject ; and,
in fact, Mr. LAWSON firſt introduced ſeveral *Engliſh* plants into notice. It is to him
that we owe the botanical part of ROBIN-
SON's " Eſſay on the Natural Hiſtory of
" *Weſtmoreland* and *Cumberland.*"

The very frequent mention of his name
in the writings of Mr. RAY, and the copi-
ouſneſs of the *Catalogue of Rare Plants,* diſtinguiſhed by him at ſo early a period, in
the North of *England,* ſufficiently confirm
the character that eminent naturaliſt gives
him, " of a diligent, induſtrious, and ſkilful
" botaniſt." It appears that he travelled

into various parts of *England*; fince he re-
cites plants collected by him on *Salifbury*
Plain; and, if I miftake not, he made an
excurfion into the *Ifle of Man*.

I conjecture that he furvived Mr. RAY
feveral years : but he was not living at the
publication of the third edition of the *Synopfis Stirpium*, although he left papers, of
which DILLENIUS availed himfelf on that
occafion.

DR. ROBINSON.

At this period muft alfo be introduced
Dr. TANCRED ROBINSON, phyfician,
in *London*, and fellow of the Royal College of Phyficians, and of the Royal Society, between whom and Mr. RAY there
fubfifted the moft genuine friendfhip and
affection. *Amicorum Alpha* is the diftinction which RAY gives him. The printed
corrrefpondence between them commences
during Dr. ROBINSON's travels abroad, in
1683, and is continued for upwards of ten
years. Seventeen letters of Dr. ROBIN-
SON appear in the " Philofophical Corre-
" fpondence," with all Mr. RAY's anfwers.
They

They run much on the ſubject of zoology;
but contain alſo botanical, and philoſophical obſervations.

Theſe letters, and the publications of
Dr. ROBINSON, in the *Philoſophical Tranſactions*, prove him to have been a man well
acquainted with various parts of learning;
to which he added alſo an intimate knowledge of natural hiſtory, and in which he
muſt have been very early initiated; ſince
RAY, in the Prefaces to his *Hiſtoria Plantarum*, in 1686, acknowledges, in ſtrong
terms, his obligations to him, for his care
and aſſiduity in correcting and enriching his
work; adding, that he had exerted himſelf
with a zeal that could not have been exceeded, had the work been entirely his
own.

Mr. RAY afterwards put into his hands
the manuſcript of the *Synopſis*, and renewed
his acknowledgments for the benefit it received under his inſpection

Dr. ROBINSON was the author of the
following communications, printed in the
Philoſophical Tranſactions.

An

An Account of the four firſt Volumes
of the " *Hortus Malabaricus;*" with Remarks. N° 145. N° 198. N° 200. N° 214.

A Deſcription, with a Figure, of the
Bridge of *St. Eſprit,* in *France.* N° 160.
Vol. xiv. p. 584.

On the natural Sublimation of Sulphur
from the Pyrites, and Limeſtone, at *Ætna,*
Veſuvius, and *Solfatara.* N° 169. Vol. xv.
p. 924.

Obſervations on boiling Fountains and
ſubterraneous Steams, occaſioned by viewing that of *Parone,* near *Montpelier:* with
an enumeration of many others in various
parts of the world. N° 169. and 172. p.
922. 1038. With Remarks.

On the Lake *Avernus.* Ib. 172.

On the Truffles found at *Ruſhton,* in
Northamptonſhire; with Figures. N° 204.
Vol. xvii. p. 935.

On the *Scotch* Barnacle, and *French* Ma-
creuſe. N° 172. Vol. xv. p. 1036.

An Account of *Henry Jenkins,* who lived
169 years. N° 221. Vol. xix. p. 267.

On the Northern Auroras, as obſerved

over

over *Vefuvius,* and the *Strombolo* Iflands;
with Conjectures on the Origin of them.
N° 349. Vol. xxix. p. 483.

Obfervations, made in 1683 and 1684,
about *Rome* and *Naples:* on the *Opuntia.*
Cork Trees: *Manna:* Volcanos: Palm
Trees; and other vegetable Productions
about *Naples:* Antiquities: Birds and Fifhes.
N° 349. Vol. xxix. p. 473.

DODSWORTH.

The Rev. *Matthew* DODSWORTH, whofe
refidence appears to have been in *Yorkfhire,*
is mentioned both by RAY and PLUKE-
NET, as well acquainted with *Englifh* botany. He was the firft difcoverer of at
leaft two of the *Englifh* plants, both of
which he fent to PLUKENET.

C H A P. 36.

Dale — *Anecdotes of* — *His* Pharmacologia — *His* " Antiquities of Harwich," *written by* Silas Taylor—Dale's *valuable additions to that work* — *An early affiftant to* Ray—*His papers in the* Philofophical Tranfactions.

D A L E.

*S*AMUEL DALE, of *Braintree*, in *Effex*, the friend and neighbour of Mr. RAY, eminent for his knowledge of botany; but better known as a writer on the moft important part of the fcience, its application to the purpofes of phyfic. I am not furnifhed with any anecdotes concerning this refpectable writer, further than, that he practifed as an apothecary at *Braintree*, until about the year 1730; about which time he became a Licentiate of the College of Phyficians, and was elected a Fellow of the Royal Society. At this time, I apprehend, he fettled at *Bocking*, and practifed as a phyfician until his deceafe, June 6th

1739,

1739, in the eightieth year of his age. He was buried in the Diſſenter's burying-ground at *Bocking*. A print of him may be ſeen before the third edition of his *Pharmacologia*.

He publiſhed PHARMACOLOGIA, ſ. *Manuductio ad Materiam Medicam*. It was firſt printed in 8°, 1693, with the ſanction of the College of Phyſicians, and we find four editions of it printed abroad. It was re-publiſhed at London in 1705 and 1710, 8°, and a third time in 4°, in 1737, pp. 460; which edition is much improved and en-larged. The arrangement of the work is that of RAY ; and, to each chapter, throughout the vegetable kingdom, he has prefixed the characters of the genus, from the *Methodus Emendata* of that au-thor. He has moreover, with great labour, conſtructed a *Syllabus*, or ſynoptical view of all the articles under each ſection or claſs, on a more amplified plan, than that of RAY.

The *Materia Medica* of DALE, in its firſt edition, may be ſaid to have been one of the earlieſt rational books on the ſub-ject.

ject. In an interval of more than forty
years, between the first and last editions,
much of that credulity which had obtained,
respecting the powers of simples, among the
writers of the last century, had abated. Several excellent publications had taken place
abroad, which, aided by improvements at
home, enabled DALE to select better materials, and give his last edition the importance
of a new work. Scarcely in any author is
there a more copious collection of synonyms,
a circumstance which, independent of much
other intrisic worth, will long continue the
use of the book, with those who wish to
pursue the history of any article through all
former writers on the subject.

In 1730, Mr. DALE published, " The
ANTIQUITIES of HARWICH and DOVER
" COURT," in 4°, pp. 464. tab. 14. written by *Silas* TAYLOR, *Gent*. about the year
1676, with large notes, amounting to much
the greater part of the book. Howsoever
respectable our author may appear as an antiquary in this volume, he is equally so as a
naturalist, in general. His History of the
Figured Fossils of the Cliff is very exact,
and

and copious; and the view he has given, in a fhort compafs, of the various opinions relating to the origin of thefe bodies, as held by the writers of the laft, and the beginning of the prefent century, is very fatisfactory.

His *Synopfis* of the animals and vegetables of the neighbouring fea and coaft, proves his intimate and critical knowledge of his fubjects; and being embellifhed with feveral good copper-plates, renders his book a real acquifition to fcience.

It is highly probable, that, from their vicinity to each other, DALE owed to Mr. RAY, his attachment to natural hiftory, and the great proficiency he fo early made in that ftudy. We find Mr. RAY acknowledging his affiftance in collecting, and extricating the fynonyms of plants, correcting errors, and fupplying omiffions, for his *Hiftoria Plantarum,* of which the *imprimatur* bears date 1685, when DALE could not be more than 26 years old.

DALE was the author of feveral communications to the Royal Society, which

8   were

were publifhed in the *Philofophical Tranf-actions*.

A Method of making Turnip Bread: practifed in Effex in a fcarcity of corn in 1693. Printed in N° 205.

Some Obfervations on the *Vermis Aureus* of *Bartholine* (aphrodita aculeata *Lin.)* a marine animal, called a *Sea Moufe :* common on the coaft of *England*; but not having been much obferved, until Dr. *Molyneux* defcribed it, had at that time excited curio-fity. N° 249.

A Relation of two large Eels, taken on the coaft of *Effex*. One of thefe mea-fured five feet eight inches; the other feven feet, in length, They wanted the character of the *Conger*, and were judged to be frefh water eels, carried by floods into the fea. N° 238, Vol. xx. p. 90.

On feveral Infects found near *Colchefter*. N° 249.

The Cafe of a Woman, who laboured under an obftinate Jaundice, accompanied with that defect of fight, which Patholo-gifts have called *Amblyopia Crepufcularis* ;

in

in which vifion is quite loft after fun-fet, and gradually returns as day-light comes on again. Nº 211. Vol. xviii. p. 158.

Queries, relating to the *Entalia, Dentalia, Blatta byzantina, Purpura,* and *Buccina* of the Shops. Nº 197. They were anfwered by Dr. LISTER.

An Account, with figures, of three Saxon Coins, dug up at *Honedon* in *Suffolk.* Nº 205. Vol. xvii. p. 874.

An Account of *Harwich* Cliff, with a Lift of twenty-eight Species of Foffil Shells, found imbedded in the Strata. Nº 291. Vol. xxiv. p. 1568. This was much enlarged in the Notes to the Hiftory of *Harwich,* mentioned above.

An Account of MSS. left by Mr. RAY. Nº 307. Vol. xxv. p. 1282.

A Letter from *Samuel* DALE, M. L. to Sir *Hans* SLOANE, Baronet, F.R.S. containing Defcriptions of the *Moofe Deer,* and a fort of Stag in *Virginia,* with Remarks on the Flying Squirrel of *America.* Vol. xxxix. p. 384*.

* LINNÆUS applied the name *Dalea* to a new American plant of the Diadelphous clafs, communicated by MILLER

MILLER to the *Clifford Garden*, and figured in the Work that bears that name. The plant afterwards fell into the genus *Pſoralea*, eſtabliſhed by *Van* ROYEN, now comprehending many ſpecies, where it preſerves the trivial name of *Dalea*. BROWN endeavoured afterwards, to perpetuate *Dale* in his *Jamaica* Plants ; but BROWN's ſpecies became the *Eupatorium Dalea* of the *Linnæan* ſyſtem.

CHAP. 37.

Bradley — *a popular Writer on Gardening and
Agriculture—Profeſſor of Botany at* Cambridge
—Hiſtoria Plantarum Succulentarum.

Blair —*Anecdotes of—His Miſcellaneous Obſervations—Botanic Eſſays: a Book of much Inſtruc-
tion—Confirms the Doctrine of the Sexes of
Plants by Experiments*—Pharmaco-botanologia
—*Papers in the* Philoſophical Tranſactions.

BRADLEY.

R*ICHARD* BRADLEY, a popular writer on Gardening and Agriculture, in the
early part of this century, was one of the firſt
who treated theſe ſubjects in a philoſophical
manner; and, as he poſſeſſed conſiderable
botanical knowledge, is entitled to a place in
theſe Anecdotes. He firſt made himſelf
known to the public by two papers, printed
in the xxixth Volume of the *Philoſophical
Tranſactions.* One "on the Motion of the Sap
"in Vegetables *." the other, "on the quick
"Growth of Mouldineſs on Melons †." He
became a Fellow of the *Royal Society,* and

---

* Vol. xxix, p. 486.　　† Ib. p. 490.

was chosen Professor of Botany in the University of *Cambridge*, in 1724. BRADLEY was the author of more than twenty separate publications; chiefly on Gardening and Agriculture, published between the years 1716 and 1730.

His " New Improvement of Planting " and Gardening, both Philosophical and " Practical," 8°, 1717, went through repeated impressions; as did his Gentleman's " and Gardener's Kalendar," (which was the fourth part of the preceding book) both at home, and in translations abroad. His " Philosophical Account of the Works of " Nature," 4°. 1721, was a popular, instructive, and entertaining work, and continued in repute several years. The same may be said of his " General Treatise of " Husbandry and Gardening," 8°. 2 Vol. 1726; and of his " Practical Discourses " concerning the four Elements, as they " relate to the Growth of Plants." 8°. 1727. His " *Dictionarium Botanicum*." 8°. 1728, was, I believe, the first attempt of the kind in *England*.

Mr. BRADLEY was not eminent for any discoveries relating to the indigenous plants

§   of

of *England* ; but exotic botany was indebted to him for an undertaking, which there
is reafon to regret he was not enabled to
purfue and perfect. I mean his book on
Succulent Plants. As this tribe is incapa-
ble of being advantageoufly preferved in a
Hortus Siccus, there is no part of botany
that calls more effentially for a feparate
publication. His work bears the following
title, " HISTORIA PLANTARUM SUCCU-
LENTARUM, *complectens hafce infequentes
Plantas, Aloen fcilicet, Ficoiden, Cereos, Me-
locardium, aliafque ejus generis quæ in Horto
ficco coli non poffunt, fecundum Prototypum puta
naturam in tabellis æneis infculptas, earum-
dem Defcriptiones huc accedunt et Cultura.* 4°.
1716. t. 50. It was publifhed in *Decads,*
at different times, between the years 1716
and 1727 ; of which only five were completed. The whole was republifhed in 1734.
The defcriptions are in Latin and Englifh,
and the figures extremely well done in the
ftile of the time. It preferves its value, as
being cited by LINNÆUS, and as containing
fome plants not figured in any other pub-

lication.

lication.   A fpecies of *Sedum* is the only
indigenous plant contained in it.

BRADLEY gave a courfe of Lectures on
the *Materia Medica,* in *London,* in the year
1729, which he publifhed in 8°, in the fucceeding year.   He died at the latter end of
1732.

Although BRADLEY's writings do not
abound in new difcoveries, yet they are not
deftitute of interefting knowledge, collected
from contemporary gardeners, and from
books.   He was an advocate for the circulation of the fap, and made feveral new obfervations on the fexes of plants, in confequence of the production of hybrid fpecies,
by which he added ftrength to that doctrine.   He wrote inftructively on the gems
of trees; on bulbs ; on grafting; and particularly, on the methods of producing va-
riegated and double flowers.

On the whole, BRADLEY's writings,
coinciding with the growing tafte for gardening, the introduction of exotics, and
improvements in hufbandry, contributed to
excite a more philofophical view of thefe
arts,

arts, and diffuse a general and popular knowledge of them throughout the kingdom.

The industry and talents of BRADLEY were not mean ; and though unadorned by learning, were sufficient to have secured to him, that reputable degree of respect from posterity, which it will ever justly withhold from him who fails to recommend such qualifications, by integrity and propriety of conduct. In these, unhappily, Mr. BRADLEY was deficient. We learn, from the account given of him by Mr. MARTYN, that he procured the professorship in a clandestine and fraudulent manner, and afterwards neglected to perform the duties of it. The University, nevertheless, allowed him to retain the nominal distinction of Professor, and appointed Dr. MARTYN to give the lectures. Near the conclusion of his life, his conduct was so unbecoming, that it was in agitation to deprive him of this nominal title.

B L A I R.

Dr. *Patrick* BLAIR was a native of *Scot-land,* and practifed phyfic and furgery at *Dundee;* where he made himfelf firft known as an anatomift, by the diffection of an elephant, which died near that place, in 1706. He was a Nonjuror, and fo far attached to the exiled family, as to have been imprifoned in the rebellion in 1715, as a fufpected perfon. He afterwards removed to *London,* where he recommended himfelf to the *Royal Society,* by fome Difcourfes on the Sexes of Flowers. His ftay in *London* was not long; he quitted it, and fettled at *Bofton,* in *Lincolnfhire;* where, if I miftake not, he practifed phyfic during the remainder of his life. I am not able to afcertain the time of his deceafe; but I conjecture that it was foon after the publication of the Seventh *Decad* of his *Pharmaco-botanologia,* in 1728.

Dr. BLAIR's firft publication was intituled " Mifcellaneous Obfervations in Phy-" fic, Anatomy, Surgery, and Botanicks." 8°, 1718. In the botanical part of this

work,

work, he infinuates fome doubts relating to
the method fuggefted by PETIVER, and
others, of deducing the qualities of vegetables, from the agreement in natural characters; and inftances the *Cynogloffum*, as
tending to prove the fallacy of this rule.
He relates feveral inftances of the poifonous
effects of plants; and thinks the *Echium
Marinum (Pulmonaria Maritima* Lin.)
fhould be ranked in the genus *Cynogloffum*,
fince it poffeffes a narcotic power. He
defcribes, and figures, feveral of the more
rare *Britifh* plants, which he had difcovered in a tour made into *Wales*. e. g. The
*Rumex Digynus : Lobelia Dortmanna : Alif-
ma Ranunculoides : Pyrola Rotundifolia : Al-
chemilla Alpina,* &c.

But the work by which Dr. BLAIR rendered the greateft fervice to botany, origi-
nated with his " Difcourfe on the Sexes of
" Plants," read before the *Royal Society,*
and afterwards greatly amplified, and publifhed, at the requeft of feveral members of
that body, under the title of,

BOTANICK ESSAYS. 1720. 8°. pp. 414.
with four copper-plates. This treatife is

K 4   divided

divided into two parts, containing five effays. The three firft, concerning what is
proper to plants; the two laft, what is common to plants and animals.

Effay I. On the Structure of Flowers. The
Diftinction and Definition of the feveral Parts.

Effay II. Definition of the Fruit, and the
feveral Kinds.

Effay III. Of the different Methods of dif-
pofing Plants. Analyfis of the feveral Methods of Claffification, with
Critical Remarks on each.

Effay IV. On the Generation of Plants.
The Concurrence of Sexes neceffary.
Variety of Reafons in Favour of this
Doctrine. The feveral Opinions re-
lating to the Nature and Ufe of the
Farina. Mr. MORLAND's Opinion
confuted.

Effay V. Of the Nourifhment of Plants.
The *Folia Seminalia*. The Vegetation of Annuals, and of Trees; and
the Structure of the Parts explained
That there is a Circulation of the Sap
in Vegetables.

Dr.

Dr. BLAIR's treatife, as far as I can find, was the firft compleat work, at leaft in the *Englifh* language, written on the fubject; and the author fhews himfelf well acquainted, in general, with all the opinions, and arguments of authors, on the matter of each effay. The value of thefe *Effays* muft not be eftimated by the meafure of modern knowledge, though even at this day, they may be read by thofe not critically verfed in the fubject, with inftruction and improvement. A view of the feveral methods then invented, cannot be feen fo connectedly exhibited in any other *Englifh* author. Dr. BLAIR ftrengthened the arguments in proof of the Sexes of Plants, by found reafoning, and fome new experiments. His reafons againft MORLAND's opinion, of the entrance of the *Farina* into the *Vafculum feminale*, and his refutation of the Lewenhoekian theory, have met with the fanction of the moderns. If his theory of vegetation, of the nourifhment of plants, and his arguments in favour of the circulation of the fap, do not meet with the approbation of the prefent age, it

muft

muſt at leaſt be granted, that they are as rational in the principle of them, as thoſe of his predeceſſors.

Pharmaco-botanologia: or, " An Alphabe- " tical and Claſſical Diſſertation on all the " *Britiſh* Indigenous and Garden Plants of " the New Diſpenſatory." Lond. 1723— 1728. 4°. The genera and ſpecies are de- ſcribed, the ſenſible qualities and medicinal powers are ſubjoined, and the pharmaceu- tical uſes.

In this work the author notices ſeveral of the more rare *Engliſh* plants, diſcovered by himſelf in the environs of *Boſton.* The work was imperfect, being carried no far- ther than the letter H.

Dr. BLAIR was the author of the fol- lowing papers in the *Philoſophical Tranſac- tions.*

The Anatomy and Oſteology of an Ele- phant, with an hiſtorical Account of that Animal. N° 326. 327. 358. Vol. xxvii. p. 53. and 117. and Vol. xxx. p. 385. This Account was alſo ſeparately publiſhed in 4°. 1711, illuſtrated with figures.

The heat of the weather when the ani- mal died, occaſioned a precipitate diſſection; but

but the rarity of the occasion added such
zeal to the anatomist, that it is matter of
surprise that Dr. BLAIR could so amply
gratify the comparative anatomist, as he has
done in this paper. He has supplied the
deficiency of some articles, and illustrated
others, from the History of the Dissection
of an Elephant which perished at *Dublin*;
published by Dr. MOULINS, in 1682.

An Account of the *Asbestus,* or *Lapis
Amianthus,* found in the county of *Angus,*
in *Scotland.* N° 333. Vol. xxvii. p. 434.

A Dissection of a Child emaciated. N°
353. Vol. xxx. p. 631. At five months
old the child weighed only five pounds. Dr.
BLAIR could find no vestige of the *omen-
tum,* and queries whether this atrophy might
not originate in the want of that membrane.
The absence of this part was probably the
consequence, rather than the cause of this
infant's disease.

An Account of a Boy who lived a considerable Time without Food. N° 364.
Vol. xxxi. p. 28.

A Method of discovering the Virtues of
Plants by their external Structure. N° 364.
Vol.

Vol. xxxi. p. 30. Dr. BLAIR thinks it
probable that even the ancients were led, in
many inftances, by the comparifon of the
habit, to afcribe fimilar virtues to plants;
and, in others, by the conformity in the
fenfible qualities of tafte and fmell.

Obfervations on the Generation of Plants.
N° 369. Vol. xxxi. p. 216. An Experiment by Mr. *Philip* MILLER, who, on fepa-
rating the Male Spinach from the Female,
found that the Seeds ripened; but on being
fown, did not vegetate. Inftances of Hybrid
Productions among Savoy and other Cab-
bage Plants. Obfervations on Variegations
in Plants: on the Impregnation of Flowers,
by the Bees and other Infects carrying the
Farina from Flower to Flower *.

---

* HOUSTON denominated an *American* plant, defcribed
by SLOANE as a *Scorodonia*, after the name of BLAIR.
This proving to be a fpecies of *Verbena*, LINNÆUS, fenfible of the praife due to BLAIR, transferred the appellation to a *Tetrandrous* plant brought from the *Cape of Good
Hope*, nearly allied in habit to the *Heath* genus, and called
it BLÆRIA.

CHAP. 38.

Sherard—*Some account of*—*Makes several tours on the continent* — *Communications to* Ray— *Supposed author of* Schola Botanica—*Editor of* Herman's Paradisus — *Consul at* Smyrna — *Communicates the* Monumenta Teia *and* Sigean Inscription *to* Chishull — *Garden near* Smyrna — *Brings* Dillenius *into* England —*His* Pinax —Herbarium—*Endows the professorship at* Oxford.

James Sherard — *Brother to the Consul* — *Well versed in* English *Botany* —*His garden at* Eltham — *Inscription on his monument.*

DR. WILLIAM SHERARD.

*W*ILLIAM SHERARD, or *Sherwood,* the son of *George Sherwood,* of *Bushby,* in *Leicestershire,* was born in 1659, and educated at Merchant Taylors' School, till he was entered at *St. John's* College, *Oxford,* in the year 1677. Of this college he became a Fellow, and took the degree of Bachelor of Law, Dec. 11, 1683. After this time, he accompanied *Lord Viscount Townshend* in his travels; and discharged

his

his truft with fo much reputation, that he was prevailed on to take the charge of *Wriothefly*, grandfon of *William*, firft *Duke of* BEDFORD; and made a fecond tour to the continent, with equal fatisfaction to the noble family who confided in him.

He returned from this tour, as I conjecture, about the year 1693; and communicated to Mr. RAY a Catalogue of Plants, which he had remarked on Mount *Jura*, *Saleve*, and the neighbourhood of *Geneva*. This was publifhed as a Supplement in RAY's " *Sylloge Stirpium Europæarum.*" About this time we find he was in *Ireland*, with his friend Sir *Arthur* RAWDON, at *Moyra*; of whom mention has been made in the article SLOANE.

In travelling, SHERARD gratified his favourite paffion, and formed connections with the moft celebrated characters on the continent, HERMAN, BOERHAAVE, and TOURNEFORT. He was very early fkilled in *Englifh* botany; and although his publications are few, there is no doubt that he had beftowed great affiduity in the ftudy of *Englifh* plants. Need I allege any farther evidence,

evidence, than the obligations, already mentioned, which Mr. RAY acknowledges for affiftance in his " Hiftory of Plants." He travelled early into various parts of *England*, and was ever attentive to make difcoveries. He made the tour of the Weft as far as into *Cornwall*. He fearched the ifland of *Jerfey*, and communicated a Lift of Plants to Mr. RAY, to be inferted in the firft edition of the *Synopfis*, printed in 1690.

He is faid to have been the author of a book publifhed under the name of *Samuel* WHARTON, " *Schola Botanica; five, Catalogus Plantarum quas ab aliquot Annis in Horto Regio Parifienfi Studiofis indigitavit Jof. Pet.* TOURNEFORT." *Amft.* 1689. 12°. It was reprinted in 1691, and 1699. If indeed SHERARD was the author of this book, he muft have attended the lectures of TOURNE-FORT three feveral feafons. It contains a rude fketch of TOURNEFORT's Method of Botany, exemplified in a large catalogue of plants; among which are innumerable varieties, fome new fpecies collected by TOURNEFORT himfelf in the *Pyrenæan* Mountains,

Mountains, and others introduced by the care of M. FAGON.

It is to SHERARD alfo, that the learned owe the publication of HERMAN's " *Paradifus Batavus, continens plus centum Plantas affabre Ære incifas, et Defcriptionibus illuftratas.*" 4°. *Lugd. Bat.* 1698. He wrote a preface to this work, in which he relates the difficulties he met with, in reducing the author's papers into method; and which contains an account of other works of HERMAN. This preface is dated from *Geneva,* in April 1697; at which time, I apprehend, SHERARD was on his third tour, on the continent.

In the year 1700, Mr. SHERARD communicated to the Royal Society a Method of making feveral *China* Varnifhes, which were fent from the Jefuits in *China* to the Great Duke of *Tufcany.* It was publifhed in the *Philofophical Tranfactions,* N° 262. Vol. xxii. p. 525. And the next year he communicated to the Society a paper from Dr. J. DEL PASSA, on the poifonous Effects of the *Indian* Varnifh on the human Skin; which on the naked Skin of Poultry

proved

proved quite harmlefs. How foon after this time he was engaged in any public employment, I cannot determine: but, in 1702, he was one of the commiffioners for fick and wounded feamen at *Portfmouth*; and, I believe, was foon after appointed conful at *Smyrna*; a department, which, it is probable, his defire of inveftigating the plants of the Eaft had no fmall fhare in inducing him to accept. But SHERARD's knowledge and tafte was not confined to the ftudy of botany. Mr. MARTYN informs us, that, " in 1705, with *Antonio* PICENINI, he vifited the feven churches of *Afia*. In 1709 and 1716, he tranfcribed the *Monumenta Teia*, and caufed the Sigean infcription to be copied and fent to *England*; and the learned Dr. CHISHULL dedicates his account of it to him." He alfo fent an account of the ifland raifed near *Santorini*, in the *Archipelago*, on the 12th of May, 1707; which was printed in the *Philofophical Tranfactions*, N° 314. Vol. xxii. p. 67.

During his refidence at *Smyrna*, he had a country houfe at a place called *Sedekio*.

It is not yet forgotten as the refidence of
SHERARD. In 1749, HASSELQUIST vifited this retreat, and viewed, with all the
enthufiafm of a young botanift, the fpot
where " the regent of the botanic world,"
as he ftiles him, fpent his fummers, and
cultivated his garden. Here SHERARD
collected fpecimens of all the plants of *Na-
tolia* and *Greece*, and began that famous
Herbarium, which at length became the
moft extenfive that had ever been feen as
the work of one man, fince it is faid finally to have contained 12,000 fpecies. And
here he is faid to have begun the much-
celebrated *Pinax*, to which he continued to
make acceffions throughout his life. He
returned into *England*, in 1718. Soon after
which time, he had the degree of Doctor
of Laws conferred upon him by the Univerfity of *Oxford*.

In 1721, Dr. SHERARD communicated
to the Royal Society an Account of the
Poifon Wood Tree of *New England*, which
he had received from Mr. MORE. It does
not appear that the fpecies had been afcertained till Dr. SHERARD pointed it out as
the

the *Arbor Americana alatis Foliis*, &c. Pluk.
Phytogr. t. 145. f. 1. (*Rhus Vernix* Lin.)
This obfervation is printed in the *Phil,
Tranf.* N° 367. Vol. xxxi. p. 147.

In this year he returned to the continent,
and made the tour of *Holland, France*, and
Italy. Whilft at *Paris*, he found VAIL-
LANT in a declining ftate of health ; but,
anxious to preferve his papers from oblivion, VAILLANT had folicited BOER-
HAAVE to purchafe, and to publifh them.
SHERARD negociated the bufinefs, and
fpent the greateft part of the fummer with
BOERHAAVE, in reducing the manufcripts
into order. To SHERARD, therefore, principally, the learned owe the *Botanicon Pa-
rifienfe*, which was publifhed in 1727.
BOERHAAVE prefixed to this work a *Latin*
letter, written by Dr. SHERARD, giving
an account of this tranfaction ; which is
alfo more fully explained in the preface.
It was in this tour, that, being in fearch of
plants in the *Alps*, he narrowly efcaped being fhot by a peafant for a wolf.

On his return, he brought over with
him the celebrated DILLENIUS, with
whom

whom he had before correfponded, and whom he had encouraged to profecute his enquiries into the *Cryptogamia* clafs, and in publifhing his *Plantæ Giffenfes*. SHERARD had himfelf been among the earlieft in *England*, to promote attention to this hitherto neglected part of nature; and in this DILLENIUS had already excelled all who had written before him.

Although Dr. SHERARD had acquired a confiderable fortune in *Afia*, yet he lived with the greateft privacy in *London*, wholly immerfed in the ftudy of natural hiftory; except when he went to his brother's feat and fine garden at *Eltham*. Dr. DILLENIUS affifted him in his chief employment, the carrying on his *Pinax*, or Collection of all the names, which had been given by botanical writers to each plant; being a continuation of *Cafpar* BAUHINE's great plan.

Dr. SHERARD was, in a particular manner, the patron of Mr. *Mark* CATESBY; and himfelf affixed the *Latin* names to the plants of " The Natural Hiftory of *Carolina*."

He

He died Auguft 12, 1728; and, by his will, gave three thoufand pounds, to provide a falary for a profeffor of botany at *Oxford*, on condition, that Dr. DILLENIUS fhould be chofen firft profeffor. He erected the edifice at the entrance of the garden, for the ufe of the profeffor; and gave to this eftablifhment his botanical library, his *Herbarium*, and the *Pinax*.

Dr. SHERARD was among the laft of thofe ornaments in *England*, of that æra which LINNÆUS calls " the golden age of botany." Having from his earlieft years a relifh for the ftudy of natural hiftory, and in his youth acquired a knowledge of *Englifh* botany, his repeated tours to the continent, and his long refidence in the Eaft, afforded ample fcope for his improvement; and the acquifition of affluence, joined to his learning, and agreeable qualities, rendered him, after his return home, a liberal and zealous patron of the fcience, and of thofe who cultivated it *.

Some

---

* VAILLANT firft devoted the name *Sherardia* to a new genus, which was afterwards affimilated with the

Vervain.

Some manuſcripts of Dr. SHERARD's
were preſented to the Royal Society by Mr.
ELLIS, in the year 1766.

J. S H E R A R D.

James, the brother of *William* SHERARD,
was born in 1666. He praƈtiſed phyſic as
an apothecary in *London*, and was early and
ſtrongly attached to his brother's favourite
purſuit. Having become eminent and opulent in his profeſſion, he cultivated, at his
country ſeat, at *Eltham*, in *Kent*, one of the
richeſt gardens that *England* ever poſſeſſed.
It was alſo the retirement of his brother,
the conſul, after his return from *Smyrna*;
and is immortalized by the pen of DILLE-
NIUS. Mr. SHERARD is not known as an
author; but his name frequently occurs in
RAY's *Synopſis*, for his diſcoveries of rare
Engliſh plants; of which he had great
knowledge, as he is ſaid to have had of natural hiſtory in general; and his zeal for

Vervain. About the ſame time, DILLENIUS gave the
like appellation, in his *Flora Giſſenſis*, to an *Engliſh* plant
of the *Stellated* claſs, in the ſyſtem of RAY, which retains
its diſtinƈtion in the *Tetrandrous* claſs of LINNÆUS.

botany

botany was fingularly great.   To thefe he
added a relifh for the elegant and polite
arts ; and particularly for mufic, in which
he was eminently fkilled.

He inherited the bulk of his brother's
fortune ; and, in the latter part of his life,
had the degree of Doctor of Phyfic conferred upon him, if I miftake not, by the
Univerfity of *Oxford ;* and was admitted a
member of the College of Phyficians.   He
married *Sufanna,* the daughter of *Richard
Lockwood,* Efq; but died without iffue, Feb.
12, 1737, N. S. and was buried in the
church of *Evington,* near *Leicefter ;* where
his widow erected a monument to his memory, of which I infert a copy below *.
She furvived him more than four years.

* M. S.
Jacobi Sherard, M. D.
Col. Med. Lond. & Soc. Reg. Soc.
Viri multifaria doctrina cultiffimi,
In rerum naturalium Botanices imprimis fcientia,
Pene fingularis ;
Et nequid ad oblectandos amicos deeffet,
Artis muficæ peritiffimi
Accefferant illi in laudis cumulum

Mores chriſtiani, vitæ integritas
Et erga omnes
Comitas et benevolentia.
Obiit prid. id. Feb. A. D. 1737,
Annos natus 72.
Uxor Suſanna, Rich. Lockwood, arm. fil
Optimo marito
Hoc monumentum mœſtiſſima poſuit;
Quæ obiit Nov. 1741,
Et juxta maritum ſepulta eſt.

CHAP. 39.

Dillenius — *a native of* Germany — *educated at*
Gieffen — *Member of the* Academia Naturæ
Curioforum — *Account of his memoirs in the*
Mifcellanea Curiofa : *on* American *plants na-*
turalized in Europe : *coffee, &c.* — *His* Catalogus Giffenfis—*An account of that book*—*His*
numerous difcoveries in the Cryptogamia *clafs*
— Dillenius *brought into* England *by Conful*
Sherard —*Publifhes a new and greatly enlarged*
edition of Ray's Synopfis — *Employed in carry-*
ing on Sherard's Pinax.

DILLENIUS.

AFTER SHERARD, I am led in
chronological courfe, as well as by
other affociations, to a character of the
higheft worth in botanical fcience. That
harmony of tafte, and co-operation of defign, which firft connected SHERARD and
DILLENIUS, hath infeparably united their
names, as long as their works fhall endure.
DILLENIUS, though not an *Englifhman*
born, is gratefully naturalized by a nation,

to

to whofe botanical fame he gave an eminence it had not experienced from the time
of RAY. It was no mean facrifice to re-
linquifh his country, his friends, his connexions, and his profpects from a profeffion, which is, at leaft fometimes, lucrative,
that he might devote himfelf to the culture
of fcience, in a foreign land, where the extent of his views was moft probably bounded by the precarious hope of a profefforfhip
alone.

John Jacob DILLENIUS * was born at
Darmftadt, in *Germany*, in the year 168⁓.
It appears that he had his education, principally at the univerfity of *Gieffen*, a city of
Upper Heffe; and where, probably, his family had confiderable intereft and connexions; fince I find two of his contemporaries of the fame name, of whom, one was

* There is a letter extant, written by DILLENIUS, in
1727, in which he tells his correfpondent, " I had once
" a mind to have fpelled my name DILLEN, it being
" eafier to pronounce; and to make my brother do the
" fame: for my great grandfather fpelled it fo, and my
" great great grandfather DILL: but, confidering that
" my name and my father's had been fo often printed
" DILLENIUS, I have left it as it is."

a profeffor

a profeffor of medicine, and dean of the faculty of phyfic at that place; and the other,
Poliater, or public phyfician; an office, I
believe, not uncommon in *Germany*, though
unknown here; and which DILLENIUS
himfelf held in the fame city. He was very
early made a member of the *Academia Curioforum Germaniæ.* He communicated feveral papers to that fociety, which were
publifhed in their *Mifcellanea Curiofa.* The
earlieft, that I find, was a Differtation, in
the *Third Century of Obfervations,* about the
year 1715, concerning the plants of *America* which are naturalized in *Europe.* This
is a fubject which might again be taken up
by a fkilful hand, to great advantage. The
refult of obfervation, and communication
on this matter, would unqueftionably prove,
that a far greater number of plants than we
are aware of, which are now thought to be
indigenous in *Europe*, were of exotic origin.
Befides the moft obvious method, from the
garden to the dunghill, and from thence to
the field, amongft a variety of other caufes,
the importation of grain has introduced a
great number : the package of merchan-
dife,

dife, and the clearing out of fhips, have been the means of difperfing many. The *Englifh Flora*, as it now ftands, cannot contain fewer, perhaps, than fixty *acknowledged* fpecies ; and a critical examination would probably inveftigate a much greater number.

In the *Fourth Century* of the fame work, we find a critical differtation on the *(Cahve)* coffee of the *Arabians :* and on *European* coffee, or fuch as may be prepared from grain or pulfe. DILLENIUS gives the refult of his own preparations made with peafe, beans, and kidney beans ; but fays, that from *rye* comes the neareft to true coffee, and was with difficulty diftinguifhed from it.

In the *Sixth Century*, he has defcribed and figured four fpecies of dubious plants ; three of the *Spergula* genus, now *Arenariæ* ; and a *Veronica*.

In the Appendix to this *Century*, DILLENIUS gave the firft fpecimen of his accurate examination of fome plants of the *Cryptogamous* clafs ; which he afterwards purfued fo greatly to the improvement of botany.

botany. In this paper, DILLENIUS treats on the propagation of plants in general; but more particularly on that of the *Ferns,* or *capillary* plants; and of the *Moffes,* which had hitherto been confidered as deftitute of flower and feed. He defcribes the flowers of that genus, which he afterwards called *Lichenaftrum,* and which was named by MI-CHELI, *Jungermannia.* He delineates two of the *Chara* genus; fome of the *Conferva;* and feveral of the more perfect plants, particularly the *Chondrilla.* He fixed the genus *Radiola; Corrigiola,* &c. and particularly the *Centunculus;* and *Cameraria,* which was afterwards called *Montia.* To thefe he fub-joins many curious obfervations on the ufe of the *petals* and *ftamina,* all tending to confirm the doctrine of the *fexes* of plants; obferva-tions on the root of the *Equifetum;* on the duft of the *Antheræ,* and on the different fhape of that in the *Orchis,* which he fays is conical; and of that in the *Ophrys,* which is round.

In the *Ninth Century* of the fame work, he relates an experiment he made concerning

ing the Opium which he prepared himſelf, from the poppy of *European* growth.

In the *Eighth Century,* he appears as a zoologiſt, in a paper on *Leeches;* and de-ſcribes two ſpecies of the *Papilio* genus.

In 1719, he publiſhed his " Catalogue " of Plants growing in the neighbourhood " of *Gieſſen ;"* a work which eſtabliſhed his character as one of the moſt accurate botaniſts of the age.   It bears the follow-ing title:

" *Jo. Jac.* DILLENII, *M. L. Ac. Nat. Cur. Coll. Catalogus Plantarum ſponte circa Giſſam naſcentium, cum Appendice, qua, Plan-tæ poſt editum Catalogum circa et extra Giſ-ſam obſervatæ recenſentur, Specierum novarum vel dubiarum Deſcriptiones traduntur, et Ge-nera Plantarum nova, Figuris æneis illuſtrata, deſcribuntur: pro ſupplendis Inſtitutionibus Rei Herbariæ Joſephi Pitton* TOURNEFOR-TII."   *Frank. ad Mæn.* 1719. 8°. Cum tab. xvi. *Cat.* pp. 240. *App.* pp. 174. Cui ſubjicitur *Examen Reſponſionis Aug. Quir.* RIVINI.

It is dedicated to the heads of the uni-verſity of *Gieſſen ;* and contains the plants
of

of the neighbourhood, confined to a circuit of not more than a *German* mile and a half. Of this tract he has given a map in his book.

The author has prefixed " A Critical Ex- " amination of the Methods of arranging " Plants," published by RAY and KNAUT, who had founded their claffical diftinctions on the *fruit*; and of thofe published by RI- VINUS, and TOURNEFORT, founded on the *flower*. In the end, he gave the preference to RAY's fyftem, and adhered to it through- out his life. His criticifm on RIVINUS brought upon-him the refentment of the author, at that time far advanced in years, who anfwered his objections. DILLENIUS had written in a ftile that was but too re- prehenfible; and can only be excufed, in fome meafure, as natural to the warmth of a young author; though it has been gene- rally acknowledged, that he had the advan- tage in the argument.

Nothing can fhew the early fkill and in- defatigable induftry of DILLENIUS more ftrongly, than his being able to produce fo great a number of plants in fo fmall a tract.

He

6

He has not enumerated fewer than 980 species, of what were then called the more perfect plants; that is, exclusive of the *Mushroom* class, and all the *Mosses*. DIL-LENIUS entered minutely into the examination of this class; and, by his diligence and discoveries, extended the bounds of that field, which the *English* botanists had so successfully cultivated before him. More had been done in *England* in this way than in any other nation. The *Pinax* of *Caspar* BAUHINE contains but fifty species; so little had the *Musci* been regarded before. The first edition of RAY's *Synopsis*, printed in 1690, not more than about eighty kinds; whereas by the investigations of the *English* botanists, particularly of DOODY, SHERARD, VERNON, LLHWYD, ROBIN-SON, PETIVER, BOBART, and others, this order was so far augmented in the second edition of the same work, in 1696, as to contain upwards of 170 species.

DILLENIUS was, however, the first writer who examined them with a view to generical characters, and divided the *Mosses*, and *Mushrooms*, each into separate *genera*.

It

It is in this book that we firſt meet with *Bryum, Hypnum, Mnium, Sphagnum, Liche-noides,* and *Lichenaſtrum,* as generical names. The four firſt of theſe, were terms in uſe with the *Patres Botanici,* although neglected by the eſtorers of the ſcience, who had ranked all under the general term *Muſcus;* except the *Lichen, Lycopodium,* and *Polytri-chum.* To demonſtrate his accuracy and diligence, it may be obſerved, that, in the environs of *Gieſſen* alone, DILLENIUS diſ-covered more than 200 ſpecies of *Moſſes,* of which 140 were new: of the *Muſhroom* order he enumerates 160, of which up-wards of 90 were ſuch as he judged had not been noticed by any author before. The plants in this catalogue are diſpoſed in the order of flowering, throughout all the year. The places of growth are ſub-joined, with critical obſervations on many of the ſpecies.

The Appendix contains a liſt of twenty plants, additional to thoſe of the Cata-logue, diſcovered in the immediate environs of *Gieſſen;* and an enumeration of upwards of 100 ſpecies, obſerved by the author, be-

yond the bounds circumfcribed in the Catalogue. This renders the book, in a great
meafure, a *Flora* of the plants of *Heffe*.
Then follows a defcription of the *new* fpecies of the Catalogue. Thefe are fucceeded
by the eftablifhment of his new *genera* of
Moffes, Fungi, and a variety of others, a-
mounting to near 100, of which fome of
former authors are here only amended; but
the far greater part are of his own conftructing, and entirely new; and the parts
of fructification feparately delineated, in 16
copper-plates. This part of his work has
been of great authority with fucceeding
writers; and many of thefe characters have
ftood the teft of the *Linnæan* fyftem.

The merit of this work fixed the character of the author, as a perfectly fcientific botanift, and attracted the notice of all
the eminent profeffors, and admirers of the
fcience: among others, that of Mr. *William,*
afterwards Dr. SHERARD, to whom we
owe that DILLENIUS was brought to *England,* and in the end fixed in the profefforfhip at *Oxford.* SHERARD was, at that
time, among the few who patronized and
cultivated the fcience in *England.* He was
lately

lately returned from *Smyrna*; and having
regretted the neglect of the *Cryptogamia*
clafs, he was fo enamoured with the difcoveries of DILLENIUS in that branch, that
he entered into correfpondence with him,
and procured fpecimens from him, and
afterwards brought him to *England.* No
man was more clofely devoted to a favourite
mufe than DILLENIUS was to *Flora*; and,
after his arrival in *England,* he purfued his
ftudy with uncommon ardour, and corre-
fponding diligence. The acquifition of fo
able a man, was probably an additional motive with the Conful, to attempt the revival of botany in the univerfity of *Oxford.*

DILLENIUS came into *England* in Auguft 1721, where he had not long refided
before he undertook a work that was much
defired; that of publifhing a new edition
of the *Synopfis Stirpium Britannicarum* of
RAY. It had been laft printed in 1696,
and was become fcarce. DILLENIUS having firmly attached himfelf to RAY's fyftem, and even improved it in fome parts
(though he intimates in one of his letters
to a friend, that he was not allowed to

make

make all the changes he wifhed for), and
being furnifhed with ample means of en-
larging the book, by his difcovery of new
fpecies of *Cryptogamia*, and by the eftablifhment of new *genera*; being alfo enabled,
by the difcoveries of many ingenious men,
whofe names he enumerates in the preface,
greatly to enlarge *Englifh* botany; and,
through the fkill and affiftance of Dr. RI-
CHARDSON, Mr. *James* SHERARD, and
others, being fufficiently qualified to add
the old *Britifh*, if I may fo fay, to the *Englifh* botany, he publifhed a third edition in
1724, much to the fatisfaction of all the
lovers of the fcience throughout *Europe*.
Twenty-four plates of rare plants were added to this edition; and, befides many valuable notes, and emendations in the *genera*,
the addition of new fpecies was very great.
The accumulation to this book from DIL-
LENIUS's own difcoveries, and from the
communications of others, whofe names are
mentioned in the preface, particularly thofe
of Dr. SHERARD and Dr. RICHARDSON,
amounted to near 40 new *Fungi*, as many
Marine plants, upwards of 150 *Moffes*, and
considerably

confiderably above 200 other plants, which had been difcovered to be natives of *Britain,* fince the publication of the fecond edition; the whole number of *Britifh* plants being about 2200, as they ftand in this book. But here it may be obferved, that botanifts had not at that time fufficiently eftablifhed fpecific diftinctions; and this number could not ftand the teft of the *Linnæan* rules, which has fince reduced the number to fewer than 1800.

DILLENIUS feems to have divided his time, before his eftablifhment at *Oxford,* principally between the country refidence of Mr. *James* SHERARD, at *Eltham,* in *Kent*; the Conful's houfe in town; and his own lodgings, which, in the year 1728, were in *Barking Alley.*

In the year 1727, Dr. THRELKELD publifhed his *Synopfis Stirpium Hibernicarum,* in which he had introduced fome fevere ftrictures on DILLENIUS, principally levelled at the introduction of his new generical names. He alfo inveighs againft him for unneceffarily multiplying the fpecies of

plants. See the articles, *Anagallis aquatica,
Dens Leonis, Lichenoides, Muscus trichoides,
Stellaria,* &c. DILLENIUS, though dif-
pleafed with the harfh and coarfe language
of THRELKELD's criticifms, had temper
enough to forbear entering into any controverfy on this occafion. He probably did
not think THRELKELD's objections of any
force fufficient to influence men of judgment in the fcience, as the *Irifh* botanift
had but little regarded any true principles
of generical diftinction. In a letter he
wrote foon after the publication of the *Irifh
Flora,* after complaining of the groffnefs
of THRELKELD's cenfures, he informs his
correfpondent that there was but one plant
recited in the book, which was not known
before as a native of *Ireland.* This, he
adds, is the *Pfeudo-ftachys Alpina* of *Cafpar*
BAUHINE *(Stachys Alpina* Lin.); and this
he had inferted on the authority of Mr.
HEATON's manufcript.

About this time he had it in contemplation, to publifh a new edition of the *Synopfis,*
with the addition of the old *Britifh* names;
and

and the times of flowering—an article neglected in the former editions. This design was laid aside, and an *Appendix* intended, for which ample materials were in hand, received from different quarters, particularly from Dr. RICHARDSON, of *North Bierly*, in *Yorkshire*; and from Mr. BREWER, who had resided two seasons at *Bangor*, purposely to investigate, and collect the plants of *Snowdon*, and the neighbouring parts. BREWER was very successful in his researches, and sent at different times great numbers of scarce plants to DILLENIUS. This *Appendix* also miscarried. In the mean time, all these exertions were favourable to the purpose he ever had in view, of completing the *Historia Muscorum*. *Wales* was a productive source of new subjects in this way, and DILLENIUS availed himself of BREWER's researches.

Whatever might be the precise nature of his engagement with the *Consul*, it appears that DILLENIUS, being doubtful of the success of the *Oxford* scheme, had formed a design of residing some time, if not finally fettling, in *Yorkfhire*. In a letter to a correfpondent of that county, dated Dec. 16, 1727, he writes thus: " Pray Sir, how is it to board in that " country? if I have done here, and *Ox-* " *ford* fails, as its likely it may do, I could " refolve to go and live there fome time, " if not for good and all; if any fmall " bufinefs fhould encourage it." Ever fince his refidence in *England*, his employ-ments had been various, and important, and his affiduity as diftinguifhed as his abilities. Since his arrival in 1721, he had publifhed the *Synopfis*, of which he defigned, if he did not himfelf engrave, all the figures. He foon after began the *Hortus Elthamenfis*. He collected materials for a new edition of, or Appendix to, the *Synopfis*. He never loft fight of his *Hiftoria Mufcorum*. Addi-tional to all which, the bufinefs of the *Pi-nax* appears to have been purfued with vigour. In a letter dated Dec. 26, 1727, he fays, " We have entered almoft all au- " thors; but to put it in order, and to write " it fair, will require fome years ftill."

In

In Auguſt 1728, his friend and patron, *Conſul* SHERARD, died; in cohſequence of whoſe will, his eſtabliſhment at *Oxford* took place ſoon after; the univerſity waving the right of nomination, in conſideration of Dr. SHERARD's benefaction.

CHAP. 40.

Dillenius *eſtabliſhed in the profeſſorſhip of botany
at* Oxford—*Publiſhes the* Hortus Elthamenſis
—Linnæus *viſits the profeſſor at* Oxford—*Cor-
reſpondence with* Haller—*Aſſiſts Dr.* Shaw *in
arranging his* Oriental *and* African *plants—
His* Hiſtoria Muſcorum—*Meditates an hiſtory
of the* Funguſſes—*His death and charaſter.*

DILLENIUS.

DILLENIUS was now arrived at that
ſituation, which had probably been
the main objeƈt of his wiſhes; and which
he conſidered equally as the completion of
his hopes, the aſylum againſt future diſap-
pointments, and the field of all that gratification, for which his taſte and purſuits
prompted him to wiſh, and qualified him
to enjoy. Add to all this, he was placed
in the ſociety of the learned, in the com-
pleteſt ſenſe of that word, and at the foun-
tain of every information, which the ſtores
of both antient and modern erudition could
diſplay, to an inquiſitive mind.

The

The plan of the *Hortus Elthamenfis* had been laid fo early as the year 1724, immediately after publifhing the *Synopfis*; and fome of the plants were figured and defcribed before Dr. SHERARD's death. The work was now carried on with vigour, and was printed in 1732, under the following title:

" HORTUS ELTHAMENSIS, *feu Plantarum rariorum quas in Horto fuo Elthami in Cantio coluit Vir ornatiffimus et præftantiffimus Jacobus* SHERARD, *M.D. Soc. Reg. et Coll. Med. Lon. Soc. Gulielmi, P.M. Frater, Delineationes et Defcriptiones, quarum Hiftoria vel planè non vel imperfectè à Rei herbariæ Scriptoribus tradita fuit. Auctore Johanne Jacobo* DILLENIO, *M.D.*" *Lond. Fol.* pp. 437. *Tab.* 324..

In this elegant and elaborate work, of which LINNÆUS fays, " *eft opus botanicum quo abfolutius mundum non vidit,*" are defcribed and figured, with the moft circumftantial accuracy, 417 plants, all drawn and etched with his own hand, confifting principally of fuch exotics as were then rare, or had been but lately introduced into *England.* A few of the more rare *Englifh* and
Welch

Welch plants were included. They are difpofed in the alphabetical order. The figures are of the natural fize as much as may
be. The *fynonyma* of former authors are
quoted and accompanied by copious critical
examinations and obfervations, the better to
afcertain the fpecies. Several new *genera*
are eftablifhed, many of the new *Gerania*
are figured, and a very copious hiftory of
the *genus Mefembryanthemum* given; with a
fynoptical view of all the fpecies, of which
fifty-four are defcribed and figured in this
work *.

We find by the lift of graduates, that
DILLENIUS was admitted to the degree of
Doctor of Phyfic in *St. John's College*,
April 3, 1735.

In the fummer of 1736, LINNÆUS vifited the Profeffor at *Oxford*; and, although
DILLENIUS did not relifh the fexual fyftem,
about that time firft divulged, yet LIN-
NÆUS returned with the higheft opinion of

---

* The plates of the *Hortus Elthamenfis* were afterwards fold to a *Dutch* bookfeller; who caft off an impreffion, accompanied with the denominations only of the
fpecies. This was done at *Leyden* in 1774; and many
copies have found their way into this kingdom.

his

his merit; and, as I have obferved on another occafion, exprefled himfelf in thefe
terms : " *In Anglia nullus eft qui genera cu-
rat vel intelligat præterquam* DILLENIUS."

LINNÆUS, after this time, correfponded
with him, fent him his *Flora Lapponica,*
and dedicated to him the *Critica Botanica.*
On which occafion the Profeffor fent his
acknowledgments in the following terms,
in a letter, dated Aug. 18, 1737 : " *Vidi,
accepi et legi Floram tuam Lapponicam multa
cum voluptate ; utinam plures iftiufmodi nobis
proftarent tali ftudio, et cura elaboratæ, in hac
te virum præftitifti."*

During this period, DILLENIUS held frequent correfpondence and communication
with HALLER, whom he efteemed, probably the more, on account of the affinity of
his fyftem with that of RAY, which he had
himfelf adopted. It appears, that he confidered HALLER as almoft the only man
qualified to carry on the *Pinax,* and wifhed
him to have been his fucceffor.

About this time, he was employed with
Dr. SHAW, in reducing to order and afcer-
taining, that learned traveller's collection of
Oriental

Oriental plants. As they were all dried fpecimens, and the collection extenfive, confifting of 640 fpecies, it required the aid of
an able hand to diftinguifh and apply fynonyms to fo confiderable a number. This
catalogue, therefore, which is annexed, with
the engravings of a few of the plants, to
the firft edition of Dr. SHAW's elaborate
work, may be confidered eventually, as the
work of the botanical Profeffor.

After the completion of the *Hortus Elthamenfis*, he purfued his " Hiftory of
" Moffes " with great diligence. It has
been obferved before, that he had extended
his refearches into this part of nature, much
further than any preceding botanift, having
been the firft difcoverer of a great number
of fpecies, and having feparated thofe heretofore defcribed together by the general
term *Mufcus*, into feveral *genera*, under the
names of *Sphagnum, Fontinalis, Bryum,* and
Hypnum ; taking his diftinctions, as well
from the *habit* of the plant, (to which the
accurate HALLER thought he paid too much
regard,) as from the figure and fituation
of that part of the fructification which is
now

now confidered as the *capfule*. By means
of the excellent botanical library of the
SHERARDS, and free accefs to their ample
Herbarium, and that of Mr. DU BOIS, who
had, with Mr. DOODY and fevcral others,
fignalized themfelves by their difcoveries
this way fome years before, DILLENIUS
enjoyed advantages which perhaps no other
fituation could have afforded. Befides
which, to give himfelf all further opportu-
nities that *Britain* allowed of making difcoveries in this department, he took a journey himfelf into *Wales,* in the fummer of
1726. In this excurfion he was attended by
Samuel BREWER. They examined *Cader
Idris,* and took up their refidence at *Bangor ;* fearched *Snowdon, Glyder,* the *Ifle of
Anglefea ;* and vifited the *Ifle of Man.* Mr.
GREEN, a clergyman of thofe parts, was
ufeful in directing their refearches, and in
affifting DILLENIUS in the *Welch* námes
of places, and of plants. The Rev. *Little-
ton* BROWN, M. A. Fellow of the Royal
Society, is alfo commemorated, as having
communicated many fpecimens of the *Cryptogamous* tribe to our author, collected by
I   him

him in *Wales*, *Shropſhire*, and *Herefordſhire*;
and thus, by the communications of theſe,
and many other friends, whoſe aid he has
gratefully acknowledged *, he was enabled
to bring his work to that degree of perfection, which would have been impracticable
in many other ſituations. In 1741 it was
publiſhed from the *Sheldon* preſs, under the
following title:

" HISTORIA MUSCORUM, *in qua circi-
ter ſexcentæ Species veteres et novæ, ad ſuâ
Genera relatæ, deſcribuntur, et Iconibus ge-
nuinis illuſtrantur ; cum Appendice, et Indice
Synonymorum.   Opera Jo. Jac.* DILLENII,
*M.D. in Univerſitate Oxonienſi Botanices Pro-
feſſoris* SHERARDINI." 4°. 1741. pp. 552.
Tab. 85.

All the ſubjects of this volume were
drawn, and engraved with his own hand.
It comprehends all thoſe plants which

* The names of ſeveral foreigners appear in this liſt;
Dr. AMMAN, of the Imperial Academy of Sciences at
Peterſburgh; *Olaus* CELSIUS, Profeſſor of Divinity at
Upſal; Dr. *J. Frederick* GRONOVIUS, of *Leyden*; Dr.
HALLER, Profeſſor at *Gottingen*; and LINNÆUS him-
ſelf.

come

come under the name of *Mufci* and *Algæ*
in the *Cryptogamia* clafs of the fexual fyftem, except the *Fucufes*, fome of the *Ulvæ*,
Confervæ, and a very few others. The author's method is throughout as follows ;
at the head of each genus he gives the etymology of the name ; his reafons for adopting that name, and applying it to the fubject ; then the definition of his genus, followed by the fubordinate diftinctions for
the arrangement of the fpecies.

In treating on each fpecies, he gives, 1.
A new fpecific character, in terms intended
to diftinguifh it from others of the fame
genus, or fubdivifion. 2. The defcription
of the fpecies at length ; diftinguifhing alfo,
with great care, the feveral varieties ; and
referring each to the feveral figures on his
plates. 3. The general places of growth ;
and under the more rare fpecies, the particular places where they have been found,
or from whence he had received them : to
thefe is fubjoined the time when each is
found in heads, or in its moft flourifhing
ftate. 4. The fynonym of every author at
length, difpofed in chronological order ;

noticing

noticing at the same time such as are refer-
able to varieties; and frequently subjoining
a number of critical observations.   5. The
uses of particular kinds, whether in the
general œconomy of nature, or in medicine,
or the other arts and conveniences of life.
A summary view of the uses of several kinds
appears in the preface; but in the body of
the work, DILLENIUS has, with great diligence, collected numerous authorities on
these heads; which sufficiently evince, that
this almost unnoticed tribe of vegetables
hold a more considerable importance * in
the scale of utility, than a superficial view
may suggest.

When

---

* Numerous proofs of the truth of this observation occur in the various writings of modern botanists.   I refer
the reader to the *Flora Lapponica*, and *Succica*; to HAL-
LER's *Historia Stirpium Helvetiæ*; to the *Amœnitates
Academicæ*; particularly to those papers under the titles
of *Oeconomia Naturæ*, and *Usus Muscorum*. I may perhaps be allowed to refer also to a *Memoir*, which I was induced to collect some years ago on the *Lichens* alone,
which was printed in the *Philosophical Transactions*, vol.
50.   On the uses of the same genus, may also be con-
sulted *Tentamen* Historiæ *Lichenum, et præsertim Prussicorum*
of HAGEN, printed at *Koningsberg*, 1782. 8°; but above

all,

When we confider the minutenefs of the objects of his inveftigation, the accuracy of his defcriptions, the critical examination and nice difcrimination of each fpecies, the labour and fkill the author has exhibited in the felection of the *fynonyma*, and the dif-pofition of them into chronological order, which is a highly meritorious part of the plan, " The Hiftory of *Moffes* " muft be confidered as a very extraordinary perform-ance: and, notwithftanding any fubfequent improvements in the arrangement of fpecies, or in the reduction of them in confequence of more perfect obfervations, or even in the microfcopical difcoveries of HEDWIG re-fpecting the *Genera*, DILLENIUS's work muft long be the bafis of knowledge in this part of nature, and muft remain with pof-terity as an almoft unexampled inftance of patience, ingenuity, and fcience, in the au-thor. This work, moreover, poffeffes a fu-periority over every other botanical publica-

all, the *Memoires couronnés en l'année* 1786, *par l'Academie des Sciences, Belles Lettres et Arts de Lyons, fur l'Utilité des Lichens dans la Medicine et dans les Arts, par M. M.* HOFF-MAN, AMOREUX *fils, et* WILLEMET, 1787. 8°.

tion

tion that I am acquainted with, in having a complete index of the *synonyma* at length. An addition of the higheſt utility in works of this kind! and which thoſe who are converſant with the writings of LINNÆUS cannot but regret the want of, in the *Species Plantarum*.

The whole impreſſion of DILLENIUS's " Moſſes " was only 250 copies, of which 50 were on imperial paper. The original edition having become extremely ſcarce *, an impreſſion of the plates, with the names only annexed, was taken off in the year 1768, and publiſhed by *John Millan*. I here remark, that this was the firſt book printed in *England*, in which any of the *Linnæan* ſpecific charaƈters were exhibited. Both the *Flora Lapponica*, and the *Hortus Cliffortianus*, are quoted in this volume.

* Poſterity will ſcarcely believe, that at the time of the publication of this work, and during he life of the author, the demand for books of natural hiſtory was ſo ſmall in *England*, that one guinea was thought a ſufficient price for this book. At this period, ten is not deemed too much ; and, not long ſince, a copy, with the plates coloured by DILLENIUS himſelf, was ſold for twenty guineas, or upwards.

There

There is little doubt that Dillenius intended to have profecuted the *Funguſſes*, as he had done the *Moſſes*; and he appears to have had this deſign in contemplation early after he came to *England*. In a letter, written in Dec. 1726, he informs his cor-reſpondent, that " He was buſy in painting " *Fungi*;" and makes this employment an apology for not anſwering his letters in due time. We know that he correſponded with Dr. Deering on this ſubject; who was himſelf well ſkilled in the knowledge of theſe productions, and had painted a great number, ſome of which he communicated to the Profeſſor.

I have been informed, that Dr. Dille-nius was of a corpulent habit of body: this circumſtance, united to his cloſe appli-cation to ſtudy, probably tended to ſhorten his days. He was ſeized with an apoplexy in the laſt week of March, 1747; and died on the 2d of April, in the 60th year of his age.

There is a portrait of him in the pic-ture gallery, or ſchool, at *Oxford*, in which he is repreſented in the academical habit;

with this infcription—*Jacobus* DILLENIUS,
*M.D. Botanices Profeffor primus, in Acade-
mia Oxonienfi* ; but I have never heard that
any engraving was made from it *.

I have never been able to acquire that
information my curiofity hath prompted
me to wifh for, relating to the domeftic
character, habits, temper, and difpofitions
of Dr. DILLENIUS. Of thofe whom I
have converfed with, who were his contemporaries, I have learned, that he was
modeft, temperate, and gentle in all his
conduct : that he was known to few who
did not feek him ; and, as might be expected, from the bent of his ftudies, and

* The drawings, dried plants, printed books and manufcripts, &c. of DILLENIUS, came into the hands of Dr.
SEIDEL, as his executor, of whom Dr. SIBTHORP purchafed them. Among thefe are all the *Britifh Funguffes*,
drawn and painted by DILLENIUS himfelf; befides a
large collection of fuch non-defcript *Fungi*, as were difcovered fubfequent to the publication of the *Synopfis*.
Some drawings alfo of the more perfect plants, done by
DILLENIUS, but many of them unfinifhed. DILLENIUS
coloured fome copies of the *Hortus Elthamenfis* himfelf;
one of which he prefented to the *Bodleian* library. (*From
information obligingly communicated by Dr. John* SIBTHORP,
the prefent learned Profeffor at Oxford),

the

the clofe application he gave to them, that
his habits were of the reclufe kind. If it
be allowable to form any opinions of men
from the perufal of their letters, fome that
I have feen, written by him, would fug-
geft, that he was naturally endowed with
a placid difpofition, improved by a philofophical calmnefs of mind, which fecured
him in a confiderable degree from the effects of the incidental evils of life. I will
at leaft lay before the reader, in the note *,

* ———" For my little time, I have met with as many
" adverfities, and misfortunes, as any body; which, by
" the help of exercife, amufement; and reading fome of
" the Stoic philofophers, I have overcome ; and am re-
" folved that nothing fhall afflict me more. Many
" things here, as well as at my home, that hath happened
" to me, would cut down almoft any body. But two
" days ago I had a letter, acquainting me with a very
" near relation's death, whom I was obliged to affift with
" money in his calamities, in order to fet him up again
" in his bufinefs ; and now this is all gone, and there
" is fomething more for me to pay, and which is not a
" little for *me*; but it does not at all affect me. I rather
" thank God that it is not worfe. This is only one, and
" I have had harder ftrokes than this, and there lies ftill
" fome upon me. Feb. 13, 1728."

N 4   a tranfcript

a tranfcript from one of his letters, written
to a friend, labouring under the preffure of
adverfe fortune; which feems to confirm
this idea *.

* If in the commemoration of celebrated men, by the
application of their names to new *genera*, any comparative dignity or fymbolical allufion, was ever to be obferved, it became in the higheft degree decent, that to
DILLENIUS fhould be appropriated one of the moft
fplendid of the vegetable race. LINNÆUS had unqueftionably this analogy in view, when he gave to this illuftrious botanift the *Syalita* of the " Malabar Garden ;" a
Polyandrous Tree, diftinguifhed for its beautiful large
flowers and fine fruit, and not lefs for its confiderable ufe
in medicinal and œconomical purpofes.

C H A P.

C H A P. 41.

Dr. Richardſon—*the correſpondent of* Sloane *and of* Dillenius—*a diligent inveſtigator of* Engliſh *plants—Communications to the* Royal Society.
Brewer—*the aſſiſtant of* Dillenius *in his* Welch *tour.*
Harriſon—*his* Herbarium *of* 4000 *ſpecimens.*
Cole—*another aſſiſtant and correſpondent of* Dillenius — *makes a collection of* Engliſh *plants, and burns it.*

RICHARDSON.

AMONG thoſe whom DILLENIUS has recorded in the preface to the third edition of RAY's *Synopſis,* and in his *Hiſtoria Muſcorum,* as having amplified *Engliſh* botany, the names of the SHERARDS, and of Dr. RICHARDSON, obtain a ſuperior diſtinction. The merit of Dr. RICHARDSON, both from his undoubted ſkill in the ſcience, and his well known patronage of thoſe who cheriſhed it, demand a more particular commemoration than I am able to give;

ſince

fince I am unacquainted with any further circumftances relating to him, than that he was educated a phyfician, and lived at *North Bierly*, in *Yorkfhire*. There he refided upon his own eftate, which was ample enough to render the practice of phyfic totally unneceffary to his well-being, from any lucrative views. He had travelled into various parts of *England*, for the inveftigation of plants, and had been fuccefsful in his tour into *Wales*, having more efpecially made difcoveries in the *Cryptogamia* clafs. His garden was well ftored with exotics, and with a curious collection of *Englifh* plants. He was happily fituated to favour his poffeffion of the latter, with which his ftore was replenifhed from time to time by the affiftance of *Samuel* BREWER, and *Thomas* KNOWLTON, both inftances of ftrong attachment to botanical purfuits, and both refident in the fame county.

Dr. RICHARDSON lived in intimacy and correfpondence with Sir *Hans* SLOANE, Dr. DILLENIUS, and other celebrated botanifts of his time. I do not find that he publifhed on his favourite amufement; but his

name

name occurs in the *Philofophical Tranfac-tions*, as author of the following papers.

On fubterraneous Trees, or Foffil Wood, found at *Youlé*, near *York*. Vol. xix. p. 526.

Obfervations in Natural Hiftory in *York-fhire*. A Boy who lived to feventeen years of age, without any Secretion of Urine, in whom Nature fupplied this deficiency by a conftant Diarrhœa. On the Trouts of the *Welch* Lakes; on the Ermine; the Nut-hatch; and the *Regulus Criftatus*; the *Helix Pomatia*. Vol. xxviii. p. 167.

A Relation of the Fall of a Water Spout in *Lancafhire*, which tore up the ground fe-ven feet deep, formed a deep gulph near half a mile in length, and deftroyed the furface of ten acres of land. Vol. xxx. p. 1097.

A Letter from Dr. *Richard* RICHARD-SON, F. R. S. to Sir *Hans* SLOANE, Bart. concerning the Voracioufnefs of the *Squilla Aquæ dulcis* in deftroying the young Fry of Carp and Tench in Ponds. Vol. xxxviii. p. 331.

A Cafe from Mr. *William Wright*, Sur-geon of *Bradford*, concerning a large Piece

of

of the Thigh Bone (5½ inches long) taken out, and its place supplied by a Callus.

Dr. RICHARDSON died at an advanced age, about the year 1740.

BREWER.

I reluctantly pass over the names of many others, mentioned in the *Synopsis*, whose services, although they were not writers on the subject, might justly call for respectful notice : but, not being able to produce any satisfactory or interesting anecdotes relating to them, I must content myself with referring the reader to a list of them, collected with no small pains, by the present Professor of Botany at *Cambridge*, and published in the Preface to his *Plantæ Cantabrigienses*.

Having however mentioned the name of *Samuel* BREWER, his connection with DILLENIUS will not allow me to refuse a proper tribute to his memory; since his passion for *English* botany, and his skill and assiduity, enabled him to afford singular assistance to the Professor, particularly in the subjects for his " History of *Mosses*;" as in

some

some instances he had done in the *Synop-sis,* for the plants of *Mendip* and *Chedder Rocks.*

He was originally of *Trowbridge,* in *Wilts,* in which county he had a small estate. He was engaged at one time in the woollen manufactory of that place; but, I believe, proved unsuccessful in business. He attended DILLENIUS into *Wales, Anglesey,* and the *Isle of Man,* in the summer of 1726; and afterwards remained the winter, and the greater part of the next year, in that country; making his residence at *Bangor,* and taking his excursions to *Snowdon* and elsewhere, often accompanied by the Rev. Mr. GREEN, and Mr. *William* JONES. While in *Wales,* it was intended that he should have gone over to *Ireland,* to make a botanical tour through that kingdom; but that expedition never took place. So long a residence gave him an opportunity not only of seeing the beauties of summer plants, but of collecting the *Cryptogamia* in winter, when they flourish most. Here he received instructions from the Professor, collected specimens of every thing

rare,

rare, or unknown to him before, and fent them to DILLENIUS, to determine the fpecies, and fix the names. I have feen a catalogue of more than two hundred plants, many of which were ill afcertained before, all fent at one time, with the references to the *Synopfis* affixed by DILLENIUS. This journey appears to have been defigned to promote the " *Appendix* to the *Synopfis*."

In 1728, Mr. BREWER went into *Yorkfhire*, and refided, I believe, the remainder of his days at *Bradford*, in that county, in the neighbourhood of Dr. RICHARDSON, by whofe beneficence he was affifted in various ways. After his retirement into *Yorkfhire*, he meditated, and nearly finifhed, a work which was to have borne the title of " The Botanical Guide ;" but it never appeared. I cannot determine the time of his deceafe, but am affured he was living in the year 1742.

HARRISON.

At a fomewhat later period, we find the name of *Thomas* HARRISON, a tradefman at *Manchefter*, who furnifhed DILLENIUS

with

with fpecimens for his hiftory. In his
younger years he had collected a large *Her-*
barium. I have been informed by one who
infpected it in the year 1762, that it contained, at that time, near 4000 fpecimens,
including both exotic and indigenous plants.
Among the latter, the *Filices* were the
moft complete part; the other *Cryptogamia*
being but few, and the collection in general
not rich in *Britifh* fpecies. In order to
accommodate the fpecimens to the largeft
fized paper, luxuriant plants of the fmaller
kinds had been chofen; a circumftance
difadvantageous to the diftinctions of fuch
plants.

Mr. Harrison's *Herbarium* hath, I believe, fince been purchafed, at a confiderable price, and is depofited in the *Manchef-*
ter library.

COLE.

Mr. *Thomas* Cole, another of the correfpondents of Dillenius, was a diffenting
minifter at *Gloucefter*, of whom I have heard
the following anecdote : That he had collected an *Herbarium,* which, in a flight of
religious

religious zeal, and repentance, at having mifpent his time in accumulating, he committed to the flames. Mr. COLE certainly forgot, at that moment, that the key to ufeful fcience is the knowledge of things. To collect the productions of nature, in order to admire and contemplate in his works the great Author of all, is in itfelf furely not only innocent, but laudable; and, when the view is extended to the utility of man, ftill more meritorious. If the fight of Mr. COLE's collection might teach but one peafant to diftinguifh that plant, which could alleviate his own, or the affliction of his neighbour, or his friend, furely it had not been made in vain.

CHAP. 42.

Rise of Botany in Ireland — Boate — Heaton —
Silliard — Molyneux — Llhwyd *and* Sherard,
all prior to Threlkeld.
Memoirs of Threlkeld—*His* Synopsis Stirpium
Hibernicarum—*An account of that work —*
Ireland *not sufficiently examined.*
Keogh's *Herbal*—Smith's *County Histories.*

IRISH BOTANY.

*I*RELAND has been so little distinguished for the production of writers
on the subject of these sketches, that it has
not been in my power, till this late period,
to introduce to the reader's notice, any pro-
fessed work on the *Flora* of that kingdom.
The distracted state of the country, during
a great part of the last century, had doubt-
less no small share in retarding the progress
of learning and science among the *Irish.* It
does not appear, that, until the middle of
that period, any enquiries had been made
even into the natural history of the country
in general.

VOL. II.   O   *Gerard*

Gerard BOATE, a *Dutch* phyſician, began " *Ireland*'s Natural Hiſtory," which
was publiſhed by *Samuel* HARTLIB in
1652, 12°. Of this the 10th, 11th, and
12th chapters treat on Agriculture. But
the ſecond part of the work, in which the
author intended to have given the Vegetables, was never publiſhed ; if indeed it was
ever written.

There is a Mr. *Zancbo* SILLIARD, an
apothecary of *Dublin*, mentioned by PAR-
KINSON, who ſeems to have poſſeſſed ſome
botanical knowledge. But the earlieſt intelligence that I can find of any real botaniſt, a native of *Ireland*, is of a Mr. HEA-
TON, a divine, who lived at *Dublin*. I
cannot collect any anecdotes of him ; but
I find his name attached, as the firſt diſco-
verer, to many plants in How's *Phytologia*,
and to ſome in MERRETT's *Pinax* ; and,
from the number and rarity of the ſubjects
recorded, he muſt have been a perſon of
conſiderable knowledge in his way. It
appears from the ſame authorities, that he
had been much in *England*, having pointed
out the natural places of many rare plants

of

of this country. He is thought to have left a manuscript on the subject, which it is conjectured was written about the year 1641, and from which THRELKELD took the *Irish* names of plants, who says, they were much more copious and exact than he could collect from any living authority. In the number of plants, it greatly exceeds any list we have extant of the old *British* names, or of those in the *Erse* tongue, among the *Highlanders.*

Towards the latter end of the century, some information was received relating to the natural history of *Ireland*, from the tour of Dr. LLHWYD, as recorded in the *Philosophical Transactions*; and Dr. *William* SHERARD, on his visits to Sir *Arthur* RAWDON, at *Moyra,* noticed many of the rare plants of that region.

Soon after this time, the establishment of the *Philosophical Society* at *Dublin* contributed to advance, among other sciences, that of natural history; and, of those who exerted themselves to promote these pursuits, were the two brothers, Dr. *William* and Dr. *Thomas* MOLYNEUX. Their

O 2 papers

papers are numerous, and are extant in the *Philosophical Transactions*.

Dr. *Thomas* MOLYNEUX was profeſſor of phyſic in the univerſity of *Dublin*, and phyſician to the ſtate, and to the army. About the beginning of this century, he drew up ſome account of the ſpontaneous vegetables of *Ireland*; which evidence, that he had applied to the ſtudy in a ſcientific manner.

He communicated his papers to Dr. THRELKELD, who incorporated ſome of them into the body of his *Synopſis*, and placed the remainder at the end. Of Dr. THRELKELD I now proceed to give ſome account.

THRELKELD.

Caleb THRELKELD, the author of the firſt treatiſe on the plants of *Ireland*, was born the 31ſt of May, 1676, at *Keiberg*, in the pariſh of *Kirkoſwald*, in *Cumberland*. In the year 1698, he commenced maſter of arts in the univerſity of *Glaſgow*, and ſoon after ſettled at *Low Huddleſceugh*, near the place of his birth, in the character of a diſſenting miniſter. He had acquired a

taſte

tafte for botany and phyfic during his refidence at *Glafgow*; and continued to make
a confiderable progrefs in thefe ftudies, infomuch, that, in 1712, he took a doctor's
degree in phyfic at *Edinburgh*; and the
next fpring, having a ftraight income, and
a large family, he removed to *Dublin*, and
fettled there in the united character of the
divine, and phyfician. Finding himfelf likely to fucceed, in little more than a year, he
fent for his family, confifting of a wife,
three fons, and three daughters. His practice as a phyfician, foon increafed, fo far as
to enable him to drop his other character
entirely, and devote himfelf wholly to phyfic. In 1727, he publifhed his "SYNOPSIS
STIRPIUM HIBERNICARUM;" and died,
after a fhort ficknefs, of a violent fever, at
his houfe in *Mark's Alley*, *Frances Street*,
April 28, 1728; and was buried in the new
burial ground belonging to *St. Patrick's*,
near *Cavan Street*; to which place his ob-
fequies were attended by a fet of children,
educated by a fociety of gentlemen, to which
inftitution he had acted as phyfician. And
my memorialift adds, that he was much re-

gretted

gretted by the poor, to whom he had been,
both as a man, and as a phyſician, a kind
benefactor.

It does not appear that Dr. THRELKELD
publiſhed any other work than the following, though he meditated a general hiſtory
of plants:

" SYNOPSIS STIRPIUM HIBERNICA-
" RUM, *alphabeticè diſpoſitarum; ſive, Com-*
" *mentatio de Plantis indigenis, præſertim*
" *Dublinenſibus inſtituta.* Being a ſhort
" Treatiſe of Native Plants, eſpecially ſuch
" as grow ſpontaneouſly in the vicinity of
" *Dublin* ; with their *Latin, Engliſh,* and
" *Iriſh* Names, and an Abridgment of their
" Virtues ; with ſeveral new Diſcoveries.
" With an Appendix of Obſervations made
" upon Plants, by Dr. MOLYNEUX, Phy-
" ſician to the State in *Ireland.* The firſt
" Eſſay of the Kind in the Kingdom of
" *Ireland. Auɛ̌tore* CALEB THRELKELD,
" M. D. *Dublin,* 1727." 8°. pp. 262.

The author, after a dedication to the
Archbiſhop of *Armagh,* and a preface,
which, though written in a quaint ſtile,
proves him to have been a man of ſome
                                    erudition

erudition in the science, enumerates all the plants he had observed in the environs of *Dublin*, and of all such as he had gained authentic intelligence, from other parts of the kingdom. He gives, first, the old *Latin* names, generally from *Caspar* BAU-HINE's *Pinax*; then the *English* name; and afterwards the *Irish*; subjoining some account of the quality of the plant, and its use in medicine, and œconomy.

He has moreover intersperfed some curious observations: to instance, under the *Betula*, or Birch Tree, he says, " The *Irish* " grammarians remark, that all the names " of the *Irish* letters, are names of trees."

Under *Brassica*, he observes, " That the " word is only the *Celtic Praisseagh* put " into a *Latin* termination; the *Latin* be- " ing no other than the *Celtic* language " cloathed with the Æolic dialect, as *Eng-* " *lish* is the *Saxon* or *Dutch* language " cloathed with *Normandy French*, as all " antiquaries will allow."

It is observable, that THRELKELD notices the good effects of the *Lythrum Salicaria*, in a dysentery: a simple since his

time

time fo ftrongly recommended by *De*
HAEN * in the fame diforder ; and in obftinate diarrhœas. He alfo fpeaks in high
terms, and from his own experience, of
the powers ufually attributed to the *Meny-
anthes trifoliata,* or Bog-bean. He quotes
from Dr. VAUGHAN a cafe of the fatal effect of the Mackenbay, or *Euphorbia Hyberna. Dr.* MOLYNEUX has obferved, that
the *Genifta fpinofa,* or Whins (Ulex euro-
peus *Lin.)* although common in other
parts of *Ireland,* is not feen in the province of *Connaught.* A fingular fact, if the
obfervation be fufficiently accurate.

In the *Appendix,* printed from the papers
of Dr. MOLYNEUX, the reader meets with
feveral curious obfervations. Among others,
an inftance of the effects of the roots of
common Henbane upon feveral perfons, who
having eaten them inftead of fkirrets, were
affected with vertiginous fymptoms, and
in one cafe a frenzy enfued, which held the
perfon two or three days. The work concludes with the *Index* of Irifh names of

* See *Rationis Medendi,* vol. i. p. 226. 357.

plants,

plants, from the manuscript supposed to have been written, as heretofore observed, by Mr. HEATON.

THRELKELD's *Flora* is not rich in the number of plants, since it does not contain more than 535 species. The author appears to have been better acquainted with the history of plants than with plants themselves; and seems not to have studied botany in a systematic way, as may be inferred from his strictures on the third edition of RAY's *Synopsis*, noticed under the article DILLENIUS.

K E O G H.

" *Botanologia Universalis Hibernica* ; or, A
" General *Irish* Herbal, calculated for this
" Kingdom ; giving an Account of the
" Herbs, Shrubs, and Trees, naturally pro-
" duced therein, in *English*, *Irish*, and *La-*
" *tin* ; with a true Description of them,
" and their Medicinal Virtues and Qualities.
" By *John* KEOGH, A.B. Chaplain to the
" Right Hon. the Lord *Kingston*. *Corke.*
" 1735." 4°.

Not

Not having feen this work, I cannot give the reader any further information relating to it.

SMITH'S HISTORIES.

In the *County Hiftories* of *Ireland*, publifhed under the direction of the Phyfico-hiftorical Society of *Dublin* by *Charles* SMITH, we meet with catalogues of the rare plants in each diftrict. Thefe lifts, however, not being drawn up with fufficient knowledge of the fubject, want that authenticity, which the critical botanift would expect, and have not greatly enlarged the botany of *Ireland*.

In that of " The antient and prefent State " of the County of *Down*," 1744, 8°, the author fpeaks of the *Savin* as indigenous— a privilege which will fcarcely be allowed to it in that kingdom ; although Dr. MO-LYNEUX, and after him THRELKELD, had recorded it. When it is recollected for what nefarious purpofes it was originally introduced into many gardens, it may readily be conjectured to be the perpetuated offspring

2   of

of original culture, in a favourable situation.

In that of " The County of *Waterford*," many very common plants, and a considerable number of the marine species. There occurs also a case, confirming the poisonous quality of the *Hemlock Dropwort*.

In that of " The County of *Cork*," 1750, 2 vols. 8°, several of the *Alpine*, and other rare plants, occur; such are the *Dryas octopetala, Sedum dasyphyllum, Euphorbia hyberna:* but what will the critical botanist say, when he sees in this list the *Androsæmum Ascyron !*

Ireland may with reason be proud to enumerate, among its choice productions of *Flora*, the *Arbutus* of *Killarney*; nevertheless, its right as an aboriginal, is with great probability of truth contested by Mr. SMITH, in his " History of the County of " *Kerry*," 1756, 8°; in which he considers it as having been introduced by the Monks of *St. Finnian*, who founded the abbey in the sixth century.

I conclude my remarks on *Irish* botany with observing, that the varied clime, the
different

different fite of the country throughout *Ireland*; its mountains, lakes, creeks, and moors, unqueftionably afford fcope to a great variety of vegetables; and the poverty of THRELKELD's *Flora* has left a rich harveft to the *Irifh* botanift: for, notwith-ftanding the confiderable time elapfed fince the publication of his book, and the lauda-ble attempts of the *Dublin* Society, I know not that *Ireland* has fince been examined by any perfon of acknowledged abilities in the fcience. What might not fuch an ad-venturer expect, from a country, which nurtures on its mountains the *Andromeda Dabœcia*, the *Dryas octopetala*, and the *Saxifraga umbrofa* of the *Alps*; and, on the borders of its enchanting lakes, the *Ar-butus Unedo* of *Greece*.

CHAP. 43.

Martyn—*Memoirs of—With* Dillenius *eſtabliſhes a botanical ſociety in* London—*Choſen Fellow of the Royal Society, and Profeſſor of Botany at* Cambridge—*Reads lectures on the* Materia Medica—*Preſents his botanical library and* Herbarium *to the Univerſity—Writings—*Tabulæ Synopticæ—Methodus Plantarum—Decades quinque—*Tranſlation from* Tournefort—*His* Virgil.

MARTYN.

AT the dawn of learning, the ſeeds of botany had been firſt ſown in *England*, by Dr. TURNER, at *Cambridge*. They can ſcarcely, however, be ſaid to have germinated, until a century afterwards, under the foſtering care of Mr. RAY. By his cultivation, they took root, although not invigorated by public ſupport. In the mean time, through the munificence of the Earl of DANBY, *Oxford* experienced the benefit of a public inſtitution in aid of this ſcience, and botany flouriſhed under the care of Mo-

RISON.

RISON. After his time, to the eftablifhment of DILLENIUS, it languifhed; no publication marked its progrefs; and its hiftory
at *Oxford* is void of interefting facts. Nearly
the fame languor prevailed after the time of
Mr. RAY at *Cambridge*, and botany attained
no ftrength till the time of Dr. MARTYN,
who, under the patronage of the univerfity,
gave the firft public lecture in that department, in the year 1727.

Of this learned botanift, I am now, in
the order of time, to prefent the reader with
fome account: and here I find myfelf agreeably anticipated by the relation of his life
and writings, prefixed to his " Differtations
on the Æneids of VIRGIL," printed in 1770,
12°, and drawn up by his moft refpectable
fon, and fucceffor in the profefforfhip; with
whofe friendfhip and correfpondence, I have
on this occafion a fincere pleafure in acknowledging, I have long been honoured.
Hence I fhall briefly recite from thefe anecdotes, only the leading circumftances in the
life of Dr. MARTYN, as connected with his
profefforial character; and conclude with a
fhort account of his botanical writings.

John

John MARTYN was born in the city of *London*, Sept. 12, 1699, and was defigned by his father for the profeffion of a merchant; but his early and ftrong propenfity to learning and fcience, in the end over-ruled that defign. He had from his youth an attachment to botany; and this tafte was further excited by his acquaintance with Mr. WILMER, afterwards demonftrator at *Chelfea* Garden; and confirmed by an intimacy with, and the countenance of, Dr. SHERARD, in the year 1719. In the year 1720, he tranflated from the *French*, Dr. TOURNEFORT's " Hiftory of the Plants " growing about *Paris*;" and having projected a like catalogue of the plants about *London*, he collected, with unwearied diligence, the native plants of the environs; making for this purpofe fometimes very extenfive excurfions, and almoft ever on foot. He had once conceived a fcheme for forming a method from the *Seed-leaves*, and had fown a great number of feeds in order to obferve the difference between them. He early became acquainted with DILLENIUS, and co-operated with him in forming a fociety

ciety of botanifts, which confifted of feventeen members. This fociety kept together
till the year 1726. He continued, during the
years 1723 and 1724, to make his excurfions in fearch of plants more frequent, and
extended them farther, into *Middlefex, Surrey, Effex,* and *Kent.* At the fame time he
ftudied Infects, continued his obfervations
on the Seed-leaves, and made many others
on the Sexes of Plants. He had, feveral
years before this time, tranflated from the
Latin, an ode on that fubject, prefented to
CAMERARIUS, and printed in that Author's
epiftle *De Sexu Plantarum.* The tranflation
may be feen in BLAIR's "Botanic Effays."

In the fummer of 1724, he travelled into
Wales, by *Bath* and *Briftol,* returning by
Hereford, Worcefter, and *Oxfora;* by which
he extended the objects of his ftudies, and
augmented his collection of *Englifh* plants;
infomuch, that at length it comprehended
1400 fpecimens.

In 1725 and 1726, he read lectures in
botany in *London,* and was recommended
by Dr. SHERARD and Sir *Hans* SLOANE
to exercife the fame function at *Cambridge;*
where,

where, on the death of BRADLEY, he was chosen Professor of Botany; and continued to give lectures for several years, until the want of a garden, and his long absence from the business of physic, which he had engaged in, rendered it incommodious to him.

In 1727, Dr. MARTYN was admitted a member of the Royal Society; and was so active in the committee for regulating the library and museum, in 1731, that he had his bond for annual payment cancelled by an order of council, as an acknowledgment of his services.

In 1730, he was admitted of *Emanuel* College, with an intention to have proceeded regularly with the degrees in physic; but his marriage, and his attention to the practice of the profession, prevented him from finishing his design. In the mean time, he read lectures in Botany and the *Materia Medica,* both at *Cambridge* and in *London,* in the years 1730 and 1731. In the beginning of the year 1733, he was elected Professor of Botany by the unanimous voice of the university.

Dr.

Dr. MARTYN had practised physic for three years in the city, but on account of an asthmatic complaint, removed in the year 1730 to *Chelsea*; where he continued the exercise of that profession, until his retirement to *Streatham*, in 1752. In 1761, he resigned his professorship; and soon after, in gratitude for the favour of having chosen him, and his son after him, to this post, he presented to the university his botanical library, consisting of upwards of 200 volumes; his *Hortus Siccus* of Exotics, containing 2600 specimens; near 250 drawings of *Fungi*; his collection of Seeds, and Seed Vessels; and his *Materia Medica*.

He removed to *Chelsea* about a year before his death; which event took place on the 29th of January, 1768.

The Professor was the author of the following publications:

TABULÆ SYNOPTICÆ *Plantarum Officinalium ad Methodum Raianam dispositæ.* 1726. fol. pp. 20. Dedicated to Sir *Hans* SLOANE.

METHODUS PLANTARUM *circa Cantabrigiam nascentium.* 1727. 12°. pp. 132.

This

This is Mr. RAY's Alphabetical Catalogue,
reduced to the order of his fyftem, with the
generic characters taken from RAY's *Me-*
thodus emendata et aucta, from VAILLANT,
DILLENIUS, SCHEUCHZER, and others,
much improved and corrected by Mr. MAR-
TYN's own obfervations. All the plants
of Mr. RAY's two *Appendices,* of 1663 and
1685, now become extremely rare, amounting to 84 fpecies, are inferted in this manual, which was printed for the ufe of his
pupils, on his firft reading lectures at *Cam-*
bridge. A fheet and an half of a new edition,
containing more than 150 fpecies, not contained in RAY's Catalogue, was printed as
part of a new edition ; but it was not carried farther : thefe were, *Submarine Plants,*
Funguſes, Moſſes, Capillaries, Apetalous and
Juliferous Plants. And, as the genius of
RAY ftill continued to animate his fucceffors, the *Cambridge Flora* has fince been
much augmented and improved, not only
by the fkill and affiduity of the prefent
Profeffor, and the labours of the late Mr.
LYONS, but more recently ftill, by the

diligent refearches and accurate difcrimina-
tions of the Rev. Mr. RELHAN.

HISTORIA PLANTARUM RARIORUM
Decades quinque. Fol. max. 1728—1732.
This was the moft fumptuous and magnificent work of the kind, that had ever been
attempted in *England*. It was dedicated
to the Royal Society, and was defigned to
contain fuch curious plants, as had not been
figured before, in their natural fize and colours; with the defcriptions, and the culture and ufes. The extraordinary expence
of this work prevented its progrefs. The
plates were mezzotinto, and printed in proper colours. Thefe Decads, among many
other rarities, contain feveral *Gerania,* the
Milleria, Martynia, Gronovia, Turnera, feveral *Paffiflora, Caffiæ,* and many *North-
American* plants.

In 1729, having entertained a defign of
reading a courfe of lectures at *Oxford,* he publifhed " The firft Lecture of a Courfe of Bo-
" tany, being an Introduction to the reft."
8°. 1729. pp. 24. tab. 84. It is an explanation of the technical terms of the fcience.

In

In the year 1720, Dr. MARTYN, as hath been obferved, had made a Tranflation of TOURNEFORT's " Hiftory of Plants about " *Paris* ;" and at the fame time meditated a Catalogue of thofe of the environs of *London.* The latter was never finifhed ; nor was the Tranflation publifhed, till twelve years afterwards, when it appeared under the following title: " TOURNEFORT's Hif- " tory of Plants growing about *Paris*, with " their Ufes in Phyfic; and a Mechanical " Account of the Operation of Medicines. " Tranflated into *Englifh*, with many Ad- " ditions, and accommodated to the Plants " growing in *Great Britain*." In 2 vols. 8°. 1732.

" The Six Alphabets" of TOURNEFORT are reduced into one; all the ufeful obfer- vations, both from the edition which came out by the united care of SHERARD and BOERHAAVE, and from that which was publifhed by *Bernard de* JUSSIEU, are ex- tracted. The Tranflator added alfo the *Englifh* names, and the places where the plants grow in *England.* He difpofed the *Moffes* according to DILLENIUS's method ;

and the *Mushrooms* and *Capillary Plants*, after a new method of his own.

Of the papers published by Dr. MAR-TYN, in the *Philosophical Transactions*, the following have relation to the subject of this work.

Rare Plants observed in a Journey into the Peak of *Derbyshire*. N° 407. Vol. xxxvi. p. 22 and 28. In this paper, the Author has taken occasion to separate the *Lactuca sylvestris murorum flore luteo* of BAU-HINE and RAY from that genus, and gives it the name of *Scariola*. LINNÆUS justifies the distinction, but calls the genus *Phrenanthes*.

An Account of a new Species of *Fungus*; N° 475. Vol. xliii. p. 263; with a Figure. Dr. MARTYN classed this singular production among the *Boleti*. He takes the opportunity, in this paper, of exhibiting a Synoptical Table of his distribution of the whole order of *Fungi :* of which it is sufficient to say, that it does not materially differ from that of DILLENIUS. The figure was copied in BLACKSTONE's *Specimen Botanicum*; and the *Fungus* has been considered

dered

dered by the author of the *Flora Anglica*, as a variety of the *Clavaria Hypoxylon* Lin.

A Remark concerning the Sex of Holly. Vol. xlviii. p. 613. Dr. MARTYN firſt obſerved the Holly Tree to be *Dioecious*, in his own garden at *Streatham*, in *Surry*. Dr. WATSON, Mr. MILLER, and ſubſequent botaniſts, not only found his obſervations true, but diſcovered, that the ſame trees bore alſo hermaphrodite flowers. This oc-caſioned the removal of it, in the *Flora Anglica*, to the claſs *Polygamia*. But as it does not appear that the remaining ſpecies of the *Ilex* are ſubjeⅽt to the ſame change, the genus ſtands in the works of LINNÆUS in the *Tetrandrous* claſs as before.

It is not without the ſtricteſt juſtice that the term indefatigable *is* applied to this learned man. His avocations from buſineſs were wholly devoted to the cauſe of litera-ture, which he contributed to ſerve in va-rious ways. The numerous works he was engaged in, and the variety of his manuſcript remains, amply teſtify this truth. At one time he was concerned in a periodical pa-per. He was a coadjutor with Mr. EAMES,

in

in abridging the *Philofophical Tranfactions*;
and was employed in writing the firft three
volumes of the " General Dictionary," in
which the lives of BELLONIUS, BOCCONE,
and BRUNSFELSIUS, were written by him.
He tranflated BOERHAAVE's " Treatife on
" the Powers of Medicine ;" HARRIS's
" Treatife on the Acute Difeafes of Infants ;"
and, jointly with Mr. CHAMBERS, gave, in
5 volumes in octavo, in 1742, a Tranflation,
or rather an " Abridgment of Philofophical
" Papers, from the Memoirs of the Royal
" Academy of Sciences at *Paris*."

Dr. MARTYN was the author of thirteen papers, printed in the *Philofophical
Tranfactions*. His Tranflation of the *Georgics* and *Bucolics* of VIRGIL, with his notes
upon this his favourite poet, hath extended
his fame among the learned of all nations.
To the claffical reader in general, they afford ample fatisfaction ; but to thofe who
join to fuch elegant enjoyment, a knowledge
of the learned Editor's favourite fcience,
thefe volumes muft afford a gratification,
which they will in vain feek for elfewhere.
His great knowledge both of antient and modern

dern fcience, relating to plants, enabled him
to appropriate the modern appellations,
with a degree of judgment, that has been
highly approved of by thofe who know the
difficulty of the undertaking, under that
almoft total want of fpecific diftinction,
which occurs in the writings of the ancients.

In the year 1737, our Author entered
into correfpondence with LINNÆUS. It
is one of thofe notices that can only occur to a lover of fimilar ftudies, that he
was, if not the firft, at leaft one of the
earlieft *Englifh* writers, who announced
the northern genius to the *Britifh* reader.
This was done by the Profeffor's extract
from the *Flora Lapponica*, printed in the
edition of the *Georgics* in 1741. It was
fome years afterwards, before the fyftem of
the *Swede* made any progrefs in *England*.

I fhall only remark further, that befides
the obligations which literature in general
owes to this learned Profeffor, that which
I call more ftrictly *Englifh* botany, received
confiderable augmentation from his labours;
particularly from his methodizing " The
" *Cambridge*

" *Cambridge* Catalogue " of Mr. RAY, and from the additions he made to his Translation of TOURNEFORT's book *.

* The name of *Martynia* was given to a fine plant of the second order in the *Diaynamous* clafs, by Dr. HOUSTON, who difcovered it on the continent of *America*. It is well known at prefent as an ornament to the *Englifh* ftoves.

C H A P. 44.

Catefby—Memoirs of—His ſtrong attachment to natural hiſtory—Reſides firſt in Virginia *ſeven years—and, encouraged afterwards by Sir Hans* Sloane *and others, returns to* America*—Natural hiſtory of* Carolina*—On birds of paſſage.*

C A T E S B Y.

ALTHOUGH the ingenious author, whom I commemorate in this chapter, does not ſtrictly rank among the improvers of indigenous botany; yet I cannot paſs over in ſilence, a man, to whom the ſcience owes one of its moſt elegant, and ſuperb productions. Mr. *Mark* CATESBY was, I believe, one of thoſe men, whom a paſſion for natural hiſtory very early allured from the intereſting purſuits of life; and it led him at length to croſs the *Atlantic,* that he might read the volume of nature in a country but imperfectly explored, and where her beauties were diſplayed in a

more

more extended and magnificent fcale, than the narrow bounds of his native country exhibited. It is but too true, that the world at large will for ever treat with ridicule and difdain that man, who, thus deferting the paths that lead to riches, to preferment, or to honour, gives himfelf up to what are commonly deemed unimportant and trifling occupations. Few will give him credit for that fecret fatisfaction, for that inexhaufti- ble pleafure, which the inveftigation of na- ture, in all her objects, inceffantly holds forth to his mind; or believe, that fuch employment can poffibly compenfate for the folid treafures of gain.

Mark CATESBY was born about the lat- ter end of 1679, or the beginning of the next year. He acquaints us himfelf, that he had very early a propenfity to the ftudy of nature; and that his wifh for higher gratifications in this way, firft led him to *London*, which he emphatically ftiles " the center of fcience;" and afterwards impelled him to feek further fources, in diftant parts of the globe. The refidence of fome re-

lations

lations in *Virginia* favoured his defign; and he went to that country in 1712, where he ftaid feven years, admiring, and collecting the various productions of the country, without having laid any direct plan for the work he afterwards accomplifhed. During this refidence, he communicated feeds and fpecimens of plants, both dried, and in a growing ftate, to Mr. DALE, of *Braintree*, in *Effex*; and, fome of his obfervations on the country, being communicated by this means to Dr. *William* SHERARD, procured him the friendfhip and patronage of that gentleman. On his return to *England*, in 1719, he was encouraged by the affiftance of feveral of the nobility, of Sir *Hans* SLOANE, Dr. SHERARD, and other naturalifts, whofe names he has recorded, to return to *America*, with the profeffed defign of defcribing, delineating, and painting the more curious objects of nature. *Carolina* was fixed on, as the place of his refidence, where he arrived in May 1722. He firft examined the lower parts of the country, making excurfions from *Charles Town*;
and

and afterwards fojourned, for fome time, a-
mong the *Indians* in the mountainous regions
at and about *Fort Moore*. He then extended his refearches through *Georgia* and *Florida*; and having fpent nearly three years
on the continent, he vifited the *Bahama
Iflands*, taking his refidence in the *Ifle of
Providence*; carrying on his plan, and particularly making collections of fifhes, and
fubmarine productions.

On his return to *England*, in the year
1726, his labours having met with the approbation of his patrons, Mr. CATESBY
made himfelf mafter of the art of Etching;
and, retiring to *Hoxton*, employed himfelf
in carrying on his great work, which he
publifhed in numbers of twenty plants each.
The firft appeared in the latter end of the
year 1730; and the firft volume, confifting
of 100 plates, was finifhed in 1732 : the
fecond, in 1743; and the Appendix, of
twenty plates, in the year 1748.

A regular account of each number, written by Dr. *Cromwell* MORTIMER, Secretary
of the Royal Society, was laid before the
Society

Society as it appeared, and printed in the
Philofophical Tranfactions; in which the
Doctor has fometimes interfperfed illuftra-
tive obfervations. *See* N° 415. 420. 426.
for Vol. i.; N° 432. 438. 441. 449. 484.
for Vol. ii.; and N° 486. for the Appendix.

The whole work bears the following
title: " The Natural Hiftory of *Carolina,*
" *Florida,* and the *Bahama Iflands*; con-
" taining the Figures of Birds, Beafts,
" Fifhes, Serpents, Infects, and Plants;
" particularly the Foreft Trees, Shrubs,
" and Plants, not hitherto defcribed, or
" very incorrectly figured by Authors; to-
" gether with their Defcriptions, in *French*
" and *Englifh*. To which are added, Ob-
" fervations on the Air, Soil, and Waters:
" With Remarks upon Agriculture, Grain,
" Pulfe, Roots. To the whole is pre-
" fixed a new and correct Map of the
" Countries treated of." By *Mark* CATES-
BY, F. R. S. *Tom.* I. 1731. pp. 100. tab.
100. *Tom.* II. 1743. pp. 100. tab. 100.
Account of *Carolina*, &c. pp. 44. Appen-

dix, tab. 20. pp. 20. *Fol. imperial,* fig.
407.

The number of fubjects defcribed and
figured in this work ftands as below :

Plants	-	-	171
Quadrupeds		-	9
Birds	-	-	111
Amphibia	-	-	33
Fifhes	-	-	46
Infects	-	-	31

In this fpendid performance, the curious
are gratified with the figures of many of the
moft beautiful trees, fhrubs, and herbaceous
plants, that adorn the gardens of the prefent time. Many alfo of the moft ufeful
in the arts, and conveniences of life, and
feveral of thofe ufed in medicine, are here
for the firft time exhibited in the true pro-
portion, and natural colours. It is only to
be regretted, that, in this work, a feparate
exhibition of the flower in all its parts fhould
be wanting; in defect of which, feveral
curious articles have not been afcertained.
It is a requifite of modern date, and with-

out

out it, every figure, efpecially of a new fpecies, muft be deemed imperfect.

Moft of the plates of plants exhibit alfo
fome fubject of the animal kingdom. To
thefe my plan does not extend ; but I will
in the note *, enumerate fome of the moft
remarkable of the vegetable clafs. As Mr.
CATESBY

* I. Of thofe ufed in food or medicine, I felect the following : The Chinkapin, *Fagus pumila* ; the nuts of
which are preferred to chefnuts, and ftored by the *Indians*
for winter food. The live Oak, *Quercus Phellos* β. of
which the acorns yield an oil not inferior to that of Almonds. The Snake-root, *Ariftolochia Virginiana* ; well
known in medicine. The May-apple, *Podophyllum pel-
tatum* ; ufed as ipecacuanha in *Carolina*. The Hiccory
tree, *Juglans alba* ; the nuts afford excellent winter pro-
vifion among the *Indians*, and yield fine oil ; the young
wood preferred for hoops, and the old for fire-wood.
The *China* root of *Carolina*, *Smilax Tamnoides*. Saffafras-
tree, *Laurus Saffafras* ; ufed in *Virginia* for intermittents.
The Cocco, and Tyre, *Arum Colocafia* ; of which the
roots are eaten by the Negroes, after deftroying the
acrimony by boiling. Ilathera Bark, *Croton Cafcarilla*.
Laurel-leaved Canella, *Canella alba* ; well known in the
fhops, and ufed as Winter's bark. The Caffena, or Ya-
pon of the *Indians*, *Prinos glaber* ; in great repute as a
reftorative. The *Virginian* Potatoe or Battatas, *Convol-
vulus Battatas* ; of general ufe as food among Whites as

CATESBY etched all the figures himself, from his own paintings, and the coloured copies were at firſt done under his own inſpection, and wherever it was poſſible, every ſubject in its natural ſize, this work was the moſt ſplendid of its kind that *England* had ever produced. I do not know that it had

well as Negroes. Marſh Cuſtard Apple, *Annona paluſtris.* *Indian* Pink, *Spigelia marilandica,* of the ſhops. Rice Plant, *Oryza ſativa.* Netted Cuſtard Apple, *Annona reticulata.* Wild Pine, or *Tillandſia polyſtachia ;* a paraſitical plant, remarkable for holding a large quantity of water in the hollow of the leaves. Mangrove Grape-tree, *Coccoloba uvifera.* Cacao, or Chocolate-tree, *Theobroma Cacao.* Vanelloe, *Epidendrum vanilla.* Caſhew Nut, *Anacardium occidentale.* Ginſeng, *Panax quinquefolium ;* the famous Ninſin of the *Chineſe.*

II. Of ſuch as more immediately reſpect the common conveniences of life, are, The Cypreſs of *America, Cupreſſus diſticha ;* the talleſt and largeſt of the *American* trees, 9 or 10 feet in diameter at the ground, and 60 or 70 high, affording a light but excellent timber. The purple Bindweed of *Carolina,* ſaid to be one of the plants the *Indians* uſe to guard againſt the venom of the Rattleſnake. The water Tupelo, *Nyſſa aquatica ;* the root ſupplies the place of corks. The Red Bay, *Laurus Borbonia ;* the wood excellent for cabinets, and beautiful as ſattin-wood. Candle-berry Myrtle, *Myrica cerifera ;* the green wax boiled

had been equalled on the continent, unless by that of Madam MERIAN, which, however, falls greatly short in extent. Seventy-two Plates of CATESBY's work were copied by the *Nuremberg* artists, and published in 1750. His " Observations on *Carolina,* &c." were separately printed in folio, at the same place, in 1767.

Mr.

boiled from the berries with one-fourth of tallow, form candles which burn long, and yield a grateful smell. Soap-wood, *Sapindus saponaria*; the bark and leaves beaten in a mortar, produces a lather used as soap. Glaucous *Mimosa*; used as sattin-wood. Brasiletto wood, *Cæsalpinia Brasiliensis*; a well-known dye. The Mangrove-tree, *Rhizophora Mangle*; forming almost impenetrable woods, the recesses of turtle, fishes, and of young alligators. The sweet Gum-tree, *Liquidambar styraciflua*; yielding a fragrant gum, like the Tolu Balsam; the wood adapted to cabinet-making. Logwood, *Hæmatoxylon campechianum.* Mahogany-tree, *Swietenia Mahagoni.*

III. Of the ornamental kind, are, The Dogwood-tree, *Cornus florida*; singular for the gradual growth of the petals, which, after the opening of the flower, expand from the breadth of a sixpence to that of a man's hand. The sweet flowering Bay, *Magnolia glauca.* The blue Trumpet-flower, *Bignonia cærulea.* Loblolly Bay, *Gordonia Lasianthus.* Carolina All spice, *Calycanthus floridus.* Tulip-tree, *Liriodendron Tulipifera.* Catalpa-tree, *Bigno-*

nia

Mr. CATESBY was the author of a paper, printed in the forty-fourth volume of the *Philosophical Transactions*, p. 435, " On Birds of Passage ;" in which, in opposition to the opinion that birds lie torpid in caverns, and at the bottom of waters, he produces a variety of reasons, and several facts,

---

nia Catolpa ; unknown in *Carolina*, till Mr. CATESBY brought it from the remoter inland parts. Sessile flowered *Trillium*. Viscous *Azalea*. Small ash leaved Trumpet-flower, *Bignonia radicans*. The Fringe-tree, *Chionanthus Virginica*. Broad-leaved Sea-side Laurel, *Xylophylla latifolia*. Willow-leaved Bay, *Laurus æstivalis*. American *Callicarpa*. Herbaceous Coral-tree, *Ærythrina herbacea*. Yellow Martagon Lily, *Lilium superbum*. *Philadelphian*, or red Martagon Lily, *Lilium Philadelphicum*. Purple *Rudbeckia*. Laurel-leaved Magnolia, *Magnolia grandiflora* ; the most superb fragrant flowering tree that ornaments our gardens. Yellow, and purple Side saddle Flower ; *Sarracenia flava, purpurea*. Umbrella Magnolia, *Magnolia tripetala*. Climbing, or four-leaved Trumpet-flower ; *Bignonia capreolata*. Lime-leaved *Hibiscus*. Red *Plumeria*. White *Plumeria*. Broad-leaved *Kalmia*. Balsam-tree, *Clusia rosea*. *Virginian* Cowslip, *Dodecatheon Meadia*. *Carolina* Pancratium. *Lilium Canadense*. Atamasco Lily, *Amaryllis atamasco*. Common *Stuartia Mulacodendron*. Blue Magnolia, *Magnolia acuminata*. *Rhododendron maximum*. And finally, the Lily-thorn, or CATESBÆA *spinosa*.

which his refidence in *America* offered, in
fupport of their migration in fearch of proper food. His voyages acrofs the *Atlantic,*
had taught him the ability of thefe wan-
derers to take long flights. He mentions,
in another place, his having feen Hawks,
Swallows, and a fpecies of *Owl*, in 26 deg.
of N. latitude, at the diftance of 600 leagues
from land. He fhews, that birds unknown
before to the country, find their way annually into various parts of *North America,*
fince the introduction of feveral kinds of
grain : of this the Rice-bird, *Emberiza ory-
zivora,* and the white-faced Duck, *Anas
difcors,* are, among others, inftances too
fufficiently known and felt by the inhabitants.

Mr. CATESBY was elected a Fellow of
the Royal Society foon after his fecond return from *America,* and lived in acquaintance and friendfhip with many of the moft
refpectable members of that body ; being
" greatly efteemed for his modefty, inge-
" nuity, and upright behaviour."

Before his death, he removed from *Hoxton* to *Fulham,* and afterwards to *London ;*

Q 3   and

and died at his houfe behind *St. Luke's* church, in *Old Street*, Dec. 23, 1749, aged 70, leaving a widow and two children*.

His work has been re-publifhed in 1754 and in 1771. To the laft edition a *Linnæan* index has been annexed; but it is by no means fo copious or perfect as a work of fuch merit and magnificence demands.

---

* Dr. GRONOVIUS called by the name of *Catefbea*, a thorny fhrub of the *Tetrandrous* clafs, bearing a long trumpet-fhaped flower, fucceeded by a yellow berry, which CATESBY firft difcovered in the *Ifle of Providence*, and fent to *Europe* in the year 1726.

C H A P.

CHAP. 45.

Houfton — *ftudied under* Boerhaave — *refident in the* Weſt Indies *for fome time — greatly aug-mented the* Chelſea Garden *with new plants— fell a facrifice to the climate — The* Reliquiæ Houſtonianæ, *publiſhed by Sir Jofeph* Banks. Douglas — *Surgeon to Queen* Caroline — *His defcription of the* Guernſey *Lily — Papers in the* Philoſophical Tranſactions.

HOUSTON.

THOSE who are converſant with the writings of MILLER, will recollect the frequent mention of the name of Dr. *William* HOUSTON; and that the exotic botany of *England* was greatly enriched by his means. If I err not, Mr. HOUSTON went firſt to the *Weſt Indies,* in the charac-ter of a ſurgeon; and, upon his return, after two years reſidence at *Leyden,* took degrees in phyſic under BOERHAAVE. This was in 1728 and 1729. At *Leyden,* he inſtitu-ted a ſet of Experiments on Brutes; ſome

Q 4

of

of which were made in concert with the late celebrated *Van* SWIETEN. They were afterwards publifhed in the *Philofophical Tranfactions*, Vol. xxxix. under the title of "*Experimenta de Perforatione Thoracis, ejufque in Refpiratione Effectibus*." The refult of which proved, contrary to the commonly received opinion, that animals could live and breathe for fome time, although air was freely admitted into both cavities of the thorax.

It appears that he was elected a Fellow of the Royal Society foon after his return from *Holland*; and that he went immediately to the *Weft Indies*. I am not able to afcertain his fixed refidence in that part of the world, although I conjecture, it was principally at the *Logwood Settlement*; from whence he fent a defcription and figure of the *Dorftenia Contrayerva*, which were publifhed in the *Philofophical Tranfactions*, Vol. xxxvii. This was the firft authentic account received of that drug, although known in *England* from the time of Sir *Francis* DRAKE, or earlier. He alfo fent to his friend at *Chelfea*, the feeds of many rare and new

new plants, collected by him in the iſlands of *Jamaica* and *Cuba*; in the province of *Venezuela,* and about *Vera Crux.*

He fell a ſacrifice to the heat of the cli-mate, and died in July 1733. He left, in manuſcript, a Catalogue of Plants, collect-ed by himſelf in the places above men-tioned; together with ſome engravings done by his own hand. Theſe came into the hands of Mr. MILLER; and, after his deceaſe, into the poſſeſſion of Sir *Joſeph* BANKS, who, out of reſpect to the memory of ſo deſerving a man, gratified the bota-niſts with the publication of them, under the following title:

" RELIQUIÆ HOUSTONIANÆ, *ſeu Plan-tarum in America meridionali, à Gulielmo* HOUSTON, *M.D. R.S.S. collectarum Icones, manu propria, ære inciſæ; cum Deſcriptioni-bus è Schedis ejuſdem in Bibliotheca Joſephi* BANKS, *Baronetti, R.S.P. aſſervatis.*" 4°. 1781. pp. 12. *tab.* xxvi.

They contain the characters and deſcrip-tions of fifteen *genera,* and eleven *ſpecies*; of which, the laſt were all natives of the country about *Vera Crux.* HOUSTON'S
new

new *genera* are defcribed in the method and
terms of Tournefort's fyftem; and all,
except one, confecrated to the memory of
botanifts; and, in this publication, they are
referred to the denominations of the *Linnæan* fyftem, as far as poffible *.

D O U G L A S.

Of the *genera* conftituted by Houston,
we find the *Douglaffia*, in honour of *James*
Douglas, F.R.S. a celebrated furgeon and
anatomift, afterwards M.D. and honorary
Fellow of the College of Phyficians, and
Phyfician to Queen *Caroline*; whom it is
juft to introduce into thefe anecdotes, fince
he obtained a reputable rank among thofe,
who in botany have been ftiled " Monogra-
" phers," from having feparately written on
a fingle fpecies or genus.   He publifhed a
very fcientific defcription of the *Amaryllis
farnienfis*, under the title of " *Lilium farni-
" enfe*; or, a Defcription of the *Guernfey*

* The name of Houstonia is given by Gronovius
to a *Tetrandrous* genus found in *Virginia*, known to the
elder authors, and fomewhat allied to the *Stellated* clafs of
Ray.

" Lily:

" Lily: to which is added, the Botanical
" Diffection of the Coffee-berry." Fol.
1725. pp. 35, and 22. *tab.* 2.

The roots of this beautiful ornament of
our prefent ftoves, were fcattered from the
wreck of a fhip on the coaft of that ifland;
and being protected, as it has been thought,
among the fand, by the Sea Reed, *Arundo
arenaria,* after the interval of fome years,
fprung up, to the furprize of the inhabitants, and the delight of the florifts and botanifts. This phenomenon will appear lefs
wonderful in our days, when it is known,
from the elegant work of Dr. THUNBERG,
that from the congeniality of climate between
England and *Japan,* one-fourth part of the
indigenous plants of that very diftant country, appear to be alfo natives of *England.*

In his " Obfervations on the Coffee,"
Dr. DOUGLAS obferves, that it was firft
mentioned by RAUWOLF in 1573, and firft
fent into *Europe* to CLUSIUS. See *Cluf.
Exotic.* p. 236.

Dr. DOUGLAS, befides many papers on
Pathological and Surgical fubjects, written
between the years 1707 and 1732, which
were

were printed in the *Philosophical Transactions*, drew up " A Botanical Description " of the *Saffron* of the Shops ;" accompanied by a figure, which was also published in the same collections, Vol. xxxii. p. 441 ; and in Vol. xxxv. the most complete account to be met with concerning the " Culture and Management of it," as practised at *Saffron Walden*. In the same volume, " An Account of the different " Kinds of *Ipecacuanha* ;" the true distinctions of which were at that time but little understood.

The knowledge of Dr. Douglas was not confined to exotic botany : he was acquainted with the plants of his own country; and his name occurs in Ray's *Synopsis*, as having noticed some rare species *.

---

* The Douglassia is lost in the *Linnæan* system, under the appellation of *Volkameria aculeata*; being an old plant of the *Didynamous* class, described by Sloane.

CHAP.

CHAP. 46.

Increasing cultivation of exotics — Superior skill of
Englifh *gardeners* — Fairchild — Knowlton —
Gordon.

Miller — *Anecdotes of — Maintained an extensive*
correspondence — His Dictionary *commended by*
Linnæus — *Member of the* Botanic Academy
at Florence—*and Fellow of the* Royal Society—
Catalogue of Hardy Trees and Shrubs—*His*
Gardener's Dictionary—Kalendar—Figures of
Plants—Cultivation of Madder—*Communica-*
tions to the Royal Society.

THE increafing cultivation of exotics
in *England,* from the beginning of the
prefent century, and the greater diffufion of
tafte for the elegancies and luxuries of the
Stove and Green-houfe, naturally tended to
raife up a fpirit of improvement and real
fcience in the arts of culture. To preferve
far-fetched rarities, it became neceffary to
fcrutinize into the true principles of the
art, which ultimately muft depend on the
knowledge

knowledge of the climate of each plant, and the foil in which it flourishes, in that climate.

Under the influence of such men as SLOANE, the SHERARDS, and other opulent encouragers of the science, gardeners acquired botanical knowledge, and were excited to greater exertions in their art. Hence, I believe, the *English* gardeners have shewn themselves equal, if not superior, to most others. My plan does not allow me to deviate so far, as to cite authors on the subject of gardening, unless eminent for their acquaintance with *English* botany. Some have distinguished themselves in this way; and I cannot omit to mention with applause, the names of FAIRCHILD, KNOWLTON, GORDON, and MILLER. The first of these made himself known to the Royal Society, by some " New Experiments re-" lating to the different, and sometimes " contrary Motion of the Sap;" which were printed in the *Philosophical Transactions*, Vol. xxxiii. p. 127. He also assisted in making experiments, by which the sexes of plants were illustrated, and the doctrine

confirmed.

confirmed. Mr. FAIRCHILD died in November 1729.

KNOWLTON.

Thomas KNOWLTON was, in the earlier part of his life, gardener to Conful SHE-RARD; but I find him in that ſtation at *Loneſborough,* in *Yorkſhire,* in the ſervice of the Earl of BURLINGTON, in the year 1728; in which place, I believe, he ſpent the greater part, if not the whole, of the remainder of his life. His zeal for *Engliſh* botany was uncommonly great, and recommended him ſuccefsfully to the learned botaniſts of this country. From Sir *Hans* SLOANE, he received eminent civilities. He merits notice in theſe memoirs, were it only to record his diſcovery of that ſingular produ&tion, the Globe *Conferva,* or Moor Balls *(Conferva Ægagropila* Lin.); which he firſt found in *Wallingfen Mere.* I have read a letter from him to a correſpondent, written in the year 1728; and another in 1729: in one of which he relates his having waded near a quarter of a mile into the lake to colle&t them; which is not done
without

without fome difficulty, as they lie at the
depth of from two to three feet.   At another time he was more fuccefsful, and collected near a bufhel at once.   He defcribes
them to his friend, under the name of *Pil-
las*, or globular Balls of Mofs, of the fize of
a tennis ball.

Mr. *Thomas* KNOWLTON was a man of
general curiofity and obfervation; and, a-
mongft other matters, not inattentive to the
purfuits of the antiquary.

We find Extracts of Two Letters from
him " to Mr. *Mark* CATESBY, F.R.S.
" concerning the Situation of the ancient
" Town *Delgovicia*, and of two Men of an
" extraordinary Bulk and Weight." *Phil.
Tranf.* Vol. xliv. p. 100.   This *Roman* ftation was difcovered on the Wolds, within
two miles of *Pocklington*.   Alfo,

" An Account of two extraordinary Deers
" Horns, found under Ground in different
" Parts of *Yorkfhire*." *Phil. Tranf.* Vol.
xliv. p. 124; with figures.   Thefe were of
two kinds: one feems to anfwer to the figure of an horn, as defcribed in *Phil. Tranf.*
N° 422. p. 257; the other was adjudged
to

to be the horns of the Moose Deer, so frequently dug up in *Ireland,* and were thought
to be the first of the kind discovered in
England.

Mr. KNOWLTON died in the year 1782,
at the advanced age of ninety.

GORDON.

James GORDON, of Mile End, eminent
for his successful cultivation of exotics, was
well acquainted with *English* botany. I
know not that he made himself known by
any publications. He maintained a correspondence with LINNÆUS; and had the
respect paid to him by the late Mr. ELLIS,
of having the *Loblolly Bay* of CATESBY
called by his name, when separated from
the *Hypericum* genus.

MILLER.

Philip MILLER was born in the year
1691. His father was gardener to the
Company of Apothecaries at *Chelsea*; and
his son succeeded him in that office, in the
year 1722. He raised himself by his merit,
from a state of obscurity, to a degree of

eminence, but rarely if ever before equalled,
in the character of a gardener. It is not
uncommon to give the term of Botaniſt,
to any man that can recite by memory, the
plants of his garden. Mr. MILLER roſe
much above this attainment. He added
to the knowledge of the theory and practice
of gardening, that of the ſtructure and characters of plants, and was early and practi-
cally verſed in the methods of RAY and
TOURNEFORT. Habituated to the uſe of
theſe, from his younger years, it was not
without reluctance that he was brought to
adopt the ſyſtem of LINNÆUS; but he was
convinced, at length, by the arguments of
the late Sir *William* WATSON and Mr.
HUDSON, and embraced it. To his ſuperior
ſkill in his art, the curious owe the culture
and preſervation of a variety of fine plants,
which, in leſs ſkilful hands, would have
failed, at that time, to adorn the conſerva-
tories of *England.*

His objects were not confined to exotics:
few were better acquainted with the indigenous plants, of which, he ſucceſſively cultivated moſt of the rare ſpecies.

2   He

He maintained a correfpondence with many of the moſt eminent botaniſts on the continent: among others, with LINNÆUS, who ſaid of his Dictionary, *Non erit Lexicon Hortulanorum, ſed Botanicorum.* By foreigners he was emphatically ſtiled *Hortulanorum Princeps.* He was admitted a member of the Botanical Academy of *Florence,* and of the Royal Society of *London,* in which he was occaſionally honoured by being choſen of the council. Mr. MILLER was the only perſon I ever knew, who remembered to have ſeen Mr. RAY. I ſhall not eaſily forget the pleaſure that enlightened his countenance, it ſo ſtrongly expreſſed the *Virgilium tantum vidi,* when, in ſpeaking of that revered man, he related to me that incident of his youth.

Mr. MILLER's infirmities induced him to reſign his office in the Garden, a little time before his deceaſe, which took place December 18, 1771, in the 80th year of his age. He left a very large *Herbarium* of Exotics, principally the produce of the *Chelſea* Garden.

In

In the year 1728, Mr. MILLER communicated to the Royal Society, " A Me-
" thod of raifing fome Exotic Seeds, which
" have been judged almoft impoffible to be
" raifed in *England.*"   *Phil. Tranf.* N° 403.
Vol. xxxv. p. 485.   This confifted in fuf-
fering the Seeds to germinate in a bark bed,
and then tranfplanting them into earth.
By this method, he fucceeded with all the
hard-fhelled fruits and feeds.   He inftances
the Cocoa Nut; the Bonduc, or Nickar
Tree *(Guilandina Bonduc* Lin.); the *Abrus
precatorius;* the Horfe Eye Bean *(Dolichos
urens)*; and feveral others.

" An Account of Bulbous Roots flower-
" ing in Bottles filled with Water." N° 418.
Vol. xxxvii. p. 81.   This method of procuring early Hyacinths, Tulips, and Nar-
ciffufes, at that time lately difcovered, is
now well known, and daily practifed.

Although he did not prefix his name to
it, he was the author of " A Catalogue of
" Trees, Shrubs, and Flowers, which are
" hardy enough to bear the cold of our cli-
" mate, and the open air ; and are propa-
" gated

" gated in the gardens near *London*." Fol.
1730. p. 90. tab. 21. The plates are coloured, the arrangement is alphabetical, and
the generical characters given. The Catalogue confifts chiefly of Trees and Shrubs;
among which are feveral of the *Coniferous*
kinds. Some varieties are interfperfed.

" CATALOGUS PLANTARUM OFFICI-
NALIUM *quæ in Horto Botanico Chelfeiano
aluntur.*" 1730. 8°. pp. 152.

In 1731, he publifhed his " Gardener's
" Dictionary," in folio, which has paffed
through many fucceffive editions; in each
of which it received fuch improvements,
and augmentations, as have rendered it in
the end the moft complete body of gardening extant. It has been tranflated into various languages; and the reception it has
every where met with, is a fufficient proof
of its fuperiority. The new edition of it,
now under the care of Profeffor MARTYN,
we doubt not, will extend to a late period,
the reputation both of the author, and of
the editor.

In the fame, or the fucceeding year, he
publifhed " The Gardener's Kalender," in 8°;

R 3   which

which has run through numerous editions, and has been a manual, in its way,
for the whole kingdom. To an edition of
this work, in 1761, the author prefixed
" A Short Introduction to the Knowledge
" of the Science of Botany ;" in which he
explains the *Linnæan* terms of art, and il-
lustrates the characters of the classes in five
copper-plates. This introduction was also
sold separately.

Mr. MILLER held an extensive correspondence with persons in distant parts of
the globe. From the *Cape of Good Hope*,
from *Siberia*, from *North America*, and particularly, by means of Dr. *William* Hous-
TON, from the *West Indies*, his garden, for
a long series of years, received a plentiful and
perpetual supply of rare, and frequently of
new species, which his successful culture sel-
dom failed to preserve. It was the remark
of foreigners, that *Chelsea* exhibited the trea-
sures of both the *Indies*. These advantages
enabled MILLER to execute, what it was
in the power of few to attempt — His
" Figures of Plants, adapted to his Dic-
" tionary," which he began to publish in
                                          numbers

numbers in 1755, and which were completed in 300 tables, making two volumes
in folio, in 1760, were drawn from plants
of his own garden. His original defign
was very extenfive; no lefs than to give
one, or more fpecies, of all the genera: but
it was found to be impracticable; and it
was therefore confined to fuch as were the
moft beautiful, ufeful, and uncommon.
Each number was accompanied with feveral pages of letter-prefs, containing the
defcriptions, and an account of the claffes
to which they belong, according to the fyftems of RAY, TOURNEFORT, and LIN-
NÆUS. As this work is well known, I fhall
only obferve, that whether we confider the
rarity of the fubjects, the fpecioufnefs of
thofe he felected for his purpofe, or the general execution of the whole, *England* had
not before produced any work, except the
Hortus Elthamenfis, and CATESBY's *Carolina,* fo fuperb and extenfive. In one refpect, MILLER's plates had the advantage
of the above mentioned, as they exhibited,
much more frequently, the feparate figures
of the parts of fructification.

R 4   " The

" The Method of cultivating Madder, as
" it is practised by the *Dutch* in *Zealand.*"
4°. 1758. Intended to excite the *English,*
by the cultivation of this important article
of trade, to fuperfede the importation of it
from the *Dutch*; who have " received from
" hence, for many years paft, more than
" 180,000 pounds a year for this root;" and
which, if properly carried on, would " em-
" ploy a great number of hands from the
" time harveft is over till the fpring, which
" is generally a dead time of the year."

" A Letter to Mr. WATSON, relating to
" a Miftake of Profeffor GMELIN, con-
" cerning the *Spondylium vulgare hirfutum.*"
C. B. *Phil. Tranf.* Vol. xlviii. p. 153.

MILLER adduces feveral reafons to
prove, that the common *Cow-Parfnep* of
Siberia, which the inhabitants make an article of food, is not the common *Cow-
Parfnep (Heracleum Spondylium)* of *Cafpar*
BAUHINE ; but the *Spondylium maximum*
of BREYNIUS : and further remarks the
miftakes that have arifen from confidering
the *common* plants of one country as the
common plants of another. On which occafion

cafion he obferves, that the *Parietaria,* fo
frequent in *England,* is not the *Parietaria
Officinarum* of *Cafpar* BAUHINE, but the
P. Ocymi folio of that author. In this fuppofition, however, we may obferve, that Mr.
MILLER has not been followed by *Englifh* botanifts of later date.

" A Letter to the Rev. *Thomas* BIRCH,
" D. D. Secretary to the Royal Society."
Phil. Tranf. Vol. xlix. p. 161. And,

" Remarks upon the Letter of Mr. *John*
" ELLIS, F. R. S. to *Philip Carteret* WEBB,
" Efq." in Vol. l. p. 430.

Thefe letters relate to a difcovery made
by the Abbé MAZEAS, and the Abbé SAU-
VAGES, on the black ftaining quality of three
feveral fpecies of *American* Sumach. Neither
the lixivium of wood afhes, nor boiling water with foap, had any effect in weakening
the tinge made by the juices of thefe plants.
They were, 1. The Poifon Afh, or *Toxicodendrum Carolinianum foliis pinnatis* (Rhus
vernix Lin.) 2. *Toxicodendron triphyllum folio
finuato pubefcente* Tourn. (Rhus Toxicodendrum). 3. *Toxicodendrum triphyllum glabrum*
(Rhus radicans). Mr. MILLER confiders the

Abbe's

Abbe's difcovery as having been long before
anticipated by KÆMPFER ; and adduces
many reafons to prove, that the *Sitzdſiu,*
or *Arbor vernicifera legitima,* p. 791. fig.
792. of that author, or the Varniſh Tree
of *Japan,* is no other than the firſt of thefe
fpecies, of which the ftaining quality is recorded by KÆMPFER. This pofition drew
Mr. MILLER into a controverfy with Mr.
ELLIS, who ftrongly infifted, that the *American* and *Japanefe Toxicodendra* were different
plants. Mr. MILLER defends his opinion
in the " Remarks." It is fufficient at this
time to obferve, that fubfequent botanifts
of the firſt note, fuch as LINÆNUS, REI-
CHARD, and THUNBERG, have counte-
nanced MILLER's opinion, by placing them
under the fame fpecific diftinction with the
Rhus vernix *.

* The MILLERIA was a new genus, difcovered at *Pa-
nama* and *Vera Cruz* by HOUSTON. It belongs to the
Syngenefious clafs, and was dedicated to MILLER by Dr.
MARTYN, in his *Decades Plantarum rariorum.*

C H A P. 47.

B L A C K W E L L.

IT is a singular fact, that physic is indebted for the most complete set of
figures of the medicinal plants, to the genius and industry of a lady, exerted on an
occasion that redounded highly to her
praise.

The name of Mrs. *Elizabeth* BLACK-
WELL is well known, both from her own
merit,

merit, and the fate of her unfortunate huſband, who, condemned for crimes of ſtate,
ſuffered death on the ſcaffold in *Sweden,* in
the year 1747.

We are informed, ſhe was the daughter
of a merchant in the neighbourhood of
Aberdeen; of which city Dr. *Alexander*
BLACKWELL, her huſband, was a native,
and where he received an univerſity education, and was early diſtinguiſhed for his
claſſical knowledge. By ſome, he is ſaid
only to have aſſumed the title of Doctor,
after his ſuccesful attendance on the King
of *Sweden*; but I believe, the more probable account is, that of his having taken
the degree of Doctor of Phyſic under BOER-
HAAVE at *Leyden*.   After having failed in
his attempt to introduce himſelf into practice, firſt in *Scotland,* and afterwards in
London, he became corrector to a printing
preſs, and ſoon after commenced printer
himſelf. But being proſecuted by the trade,
and at length involved in debt, was thrown
into priſon. To relieve theſe diſtreſſes, Mrs.
BLACKWELL, having a genius for drawing
and painting, exerted all her talents; and,
                              underſtanding

underſtanding that an Herbal of Medicinal Plants was greatly wanted, ſhe exhibited to Sir *Hans* SLOANE, Dr. MEAD, and other phyſicians, ſome ſpecimens of her art in painting plants, who approved ſo highly of them, as to encourage her to pro-ſecute a work, by the profits of which ſhe is ſaid to have procured her huſband's li-berty, after a confinement of two years.

Mr. RAND, an eminent apothecary, was at that time Demonſtrator to the Company of Apothecaries, in the Garden at *Chelſea.* By his advice ſhe took up her reſidence op-poſite the Phyſic Garden, in order to faci-litate her deſign, by receiving the plants as freſh as poſſible. He not only promoted her work with the public, but, together with Mr. *Philip* MILLER, afforded her all poſſible direction and aſſiſtance in the exe-cution of it. After ſhe had completed the drawings, ſhe engraved them on copper, and coloured the prints with her own hands.

During her abode at *Chelſea,* ſhe was fre-quently viſited by perſons of quality, and many ſcientific people, who admired her

performances,

performances, and patronized her under-taking.

On publishing the first volume, in 1737, she obtained a recommendation from Dr. MEAD, Dr. SHERARD, Mr. RAND, and others, to be prefixed to it. And being allowed to present, in person, a copy to the College of Physicians, that body made her a present, and gave her a public testimonial of their approbation; with leave to prefix it to her book. The second volume was finished in 1739, and the whole published under the following title :

" A curious Herbal, containing 500 Cuts
" of the most useful Plants which are now
" used in the Practice of Physic, engraved
" on folio copper-plates, after drawings
" taken from the life. By ELIZABETH
" BLACKWELL. To which is added, a
" short Description of the Plants, and their
" common Uses in Physic. 1739." 2 vol. fol.

The drawings are in general faithful; and if there is wanting that accuracy, which modern improvements have rendered neces-sary, in delineating the more minute parts,

yet,

yet, upon the whole, the figures are fufficiently diftinctive of the fubject.

Each plate is accompanied with an engraved page, containing the *Latin* and *Englifh* officinal names, followed by a fhort
defcription of the plant, and a fummary of
its qualities, and ufes. After thefe occur
the name in various other languages. Thefe
illuftrations were the fhare her hufband took
in the work. This ill-fated man, after his
failure in phyfic, and in printing, became
an unfuccefsful candidate for the place of
Secretary to the Society for the Encouragement of Learning. He was made Superintendant of the Works belonging to the
Duke of CHANDOS at *Cannons,* and experienced thofe difappointments incident to
projectors. He formed fchemes in agriculture, and wrote a treatife on the fubject,
which, we are told, was the caufe of his
being engaged in *Sweden.* In that kingdom, he drained marfhes, practifed phyfic,
and was even employed in that capacity for
the king. At length he was involved in
fome ftate cabals, or, as fome accounts have
it, in a plot with *Count* TESSIN, for which
he

he loft his life, protefting his innocence to the laft.

So refpectable a performance as Mrs. BLACKWELL's, attracted the attention of phyficians on the continent. TREW, of *Norimberg*, in the year 1750, engaged an artift of that place to copy Mrs. BLACK-WELL's plates, and himfelf fupplied feveral defects in the drawings. He fubftituted fome entirely new figures in the room of the originals, very confiderably reformed and amplified the text, tranflated it into *German* and *Latin*, and planned the addition of a fixth century of plates. He prefixed a moft elaborate and learned Catalogue of Botanical Authors, but did not live to finifh the work. The Fifth Century was publifhed in 1765; and Dr. TREW dying in 1769, the fupplemental volume, exhibiting plants omitted by Mrs. BLACKWELL, articles newly introduced into practice, and figures of the poifonous fpecies, was conducted by LUDWIG, BOSE, and BOEHMER, and printed in 1773. Thus reformed, TREW's edition furpaffes any other work of the fame defign. If there are imperfections in it, they

they were unavoidable, arifing from the impracticability of procuring recent fpecimens in fome inftances, and from an almoft total ignorance of the origin of others, defects ftill unfupplied in various articles.

DEERING.

Charles DEERING was a native of *Saxony*. He took his degrees in phyfic at *Leyden*; and, as Mr. MARTYN informs us, came to *England* firft, in the train of a foreign ambaffador. This happened, I conjecture, before, or about the year 1720. He practifed phyfic and midwifery in *London*; and having a ftrong bias to the ftudy of botany, became one of the members of the fociety eftablifhed by Dr. DILLENIUS and Mr. MARTYN, which fubfifted from the year 1721 to 1726.

In the year 1736, he removed to *Nottingham*, under the recommendation of Sir *Hans* SLOANE. At this time he was married; but his wife did not long furvive the removal to that place. He was at firft well received; and is faid to have been very fuccefsful in his treatment of the fmall-pox, which difeafe was highly epidemical at that place, foon after his fettling there. But he incurred the cenfure of the faculty, by his pretenfions to a noftrum. He publifhed " An Account of an improved Method of " treating the Small-pox, in a Letter to " Sir *Thomas* PARKYNS, Bart." 8°. 1737. pp. 52. By this tract it appears, that his medicine was of the antiphlogiftic kind, and his regimen the cool one, which at that time had been adopted by very few, as general practice.

Dr. DEERING fhewed his attachment to his botanical purfuits, by his affiduity in collecting fuch ample materials for his Catalogue in lefs than two years after fixing at *Nottingham*. He publifhed it under the following title:

" A CATALOGUE OF PLANTS naturally " growing and commonly cultivated in divers " Parts of *England*, more efpecially about " *Nottingham*: containing the moft known " *Latin* and *Englifh* Names of the feveral " Plants; the Tribe they belong to; the " Time of their flowering; and of thofe " which are either Officinals or otherwife,
                                                    " of

" of any known Efficacy, fuch Virtues are
" briefly mentioned as may be depended
" upon. To which is added, a general
" Diftribution of Plants according to Mr.
" RAY; with an Explanation of fome bo-
" tanical and phyfical Terms; and an al-
" phabetical Lift of Plants in Flower, for
" every Month in the Year. By *Charles*
" DEERING, *M.D. Nottingham.*" 8°.
1738. pp. 264.

The arrangement is alphabetical, and the
number of plants about 850. The author
was particularly attached to the fubjects of
the *Cryptogamia* clafs, in which his refearches had been very fuccefsful. Of the
number above mentioned, more than 200
belonged to the orders of *Fungi*, *Mufci*, and
Algæ; among which, we meet with 27
which he confidered as *nondefcripts*, and 10
others not to be met with in the third edition of RAY's *Synopfis*. He was affifted in
this branch by his correfpondence with the
learned Profeffor at *Oxford*, who confidered
fome of his difcoveries as new, and fpeaks
of his knowledge and affiduity in terms of
applaufe. In page 89 of his pofthumous

S 2   work

work, the *Nottinghamia Vetus et Nova*, there occurs a lift of fome plants, difcovered by the author after the publication of this Catalogue. Thefe are principally of the *Cryptogamous* kind. Several of the more rare plants of the environs efcaped the obfervation of this affiduous man ; among which may be mentioned particularly, that moft virulent of all our *Englifh* productions, the *Cicuta virofa*, or, long-leaved Water Hemlock ; which I remember to have feen growing in the *Leen*, near the *Rock-holes*, in *Nottingham* Park. That the *Addenda* to his " Catalogue" were not more copious may be attributed to his fubfequent misfortunes, which undoubtedly damped the ardour of his purfuit.

Notwithftanding his early fuccefs, that " adverfe fatality," which he himfelf alludes to in his " Letter on the Small-pox," ftill attended him. He was, unhappily, not endowed with that degree of prudence, and equanimity of temper, which are fo neceffary to the practice of phyfic ; infomuch, that he very early loft the little intereft which his character and fuccefs had at firft gained.

But

But as I would rather dwell on his merits, than on his failings, I fhall obferve, that befides his acquaintance with the antient languages, he was mafter of many of the modern tongues. His knowledge of that fcience which gives him a place in this fketch, was very confiderable, and will be perpetuated, fo long as DILLENIUS's " Hiftory" fhall preferve eftimation. He had a knowledge of defigning, and was an ingenious mechanic. After his failure in Phyfic, his friends attempted feveral fchemes to alleviate his neceffities. They procured him, among others, a commiffion in the regiment raifed at *Nottingham*, on account of the rebellion. But this proved more honourable than profitable to him. He was afterwards employed in a way more agreeable to his genius, and talents; being furnifhed with materials, and enabled by the affiftance of *John* PLUMTREE, Efq; and others, to write the Hiftory of *Nottingham*, which he dedicated to the *Duke of* NEWCASTLE. But he did not live to receive the reward of this labour. He had been troubled with the gout at a very early period, having been afflicted

S 3   with

with it in his nineteenth year, and in
the latter ftage of his life, he fuffered long
confinements in this difeafe, and became
afthmatical. Being at length reduced to a
degree of poverty, and dependance, which
his fpirit could not fuftain, oppreffed with
calamity, and complicated difeafe, he died
April 12, 1749. Two of his principal cre-
ditors adminiftered to his effects, and buried
him in *St. Peter's* church-yard, oppofite the
houfe in which he refided.

He left an *Hortus Siccus* of the plants of
his " Catalogue," confifting of upwards of
600 fpecies, in eight volumes, of the quarto
form; befides feparate tables of the *Moffes*,
and a volume of paintings of the *Fungi*, accurately done by his own hand. Some part,
if not the whole, of this collection, was, I
believe, purchafed by the Honourable *Roth-*
well WILLOUGHBY, who had been one of
his benefactors, while living, and inherited
a portion of that tafte, which diftinguifhed
his family in the time of Mr. RAY. He
left alfo a manufcript treatife, in *Latin, De*
Re obftetricaria.

His

His pofthumous work was publifhed by his adminiftrators, *George* AYSCOUGH, printer, and *Thomas* WILLINGTON, druggift, under the following title:

" NOTTINGHAMIA VETUS *et* NOVA : " or, An Hiftorical Account of the ancient " and prefent State of the Town of *Not-* " *tingham,* gathered from the Remains of " Antiquity, and collected from authentic " Manufcripts, and ancient as well as mo- " dern Hiftorians; adorned with beautiful " copper-plates. By *Charles* DEERING, M.D. *Nottingham.* 1751." 4°. pp. 370.

It is embellifhed with 24 copper-plates; among which are a plan, and two views of the town; a ground plan of the old caftle; two views of the prefent caftle; the three churches; and many other buildings. A view of the " Rock-holes" in the park; fuppofed by Dr. STUKELEY to have been the work of the *Britons,* enlarged and altered by the *Saxons.* But one of the moft remarkable articles in this volume is, a complete defcription of that curious machine the ftocking-frame, invented two centuries ago by *William* LEE, M.A. of *St. John's* College,

lege, *Cambridge*, a native of *Woodborough*, near *Nottingham*. I know not that so full an account of this complicated machine is elsewhere to be seen. All the parts are separately, and minutely described, in the technical terms; and illustrated by two views of the whole, and by a large table, delineating with great accuracy, every con-stituent part of the machine.

WILSON.

The subject of this article, like *Thomas* WILLISEL, and *Samuel* BREWER, is another instance of that unconquerable attachment to a favourite branch of know-ledge, which sometimes engrosses the minds of those, who, by their lot, have not been exempted from labouring in the lower, and mechanical offices of life.

From information which I received, more than twenty years ago, concerning *John* WILSON, I learned that he was originally an inhabitant of *Kendal*, in *Westmoreland*; and was employed in the manufacture of knit stockings, for which that town was so famous. That, at one time, he gave weekly

lessons

leffons on botany, alternately, at that place, and at *Newcaftle.* That many pupils reforted to him from the neighbouring parts of *Scotland;* infomuch, that in fome feafons, he received fixty pounds a year, as the premium of his labours.

I muft not, however, omit to obferve, that this account does not coincide with another, which I have fince met with in the " *Britifh* Topography;" the refpectable author of which informs us, " That WIL-" SON was a fhoemaker, and by his intenfe " application to his favourite ftudy, lived " moft of his life in a ftate of indigence. " A cow, of which his wife had the care, " was the fole fupport of his family : and " fuch was his infatuation, that he was " once tempted to part with that moft ufe-" ful animal, to purchafe MORISON's vo-" luminous work, had not a neighbouring " lady prefented him with the book, and " refcued the poor man and his family from " beggary and ruin."

In this reprefentation of WILSON's conduct, while men of fympathizing minds,
and

and fimilar tafte, muft deplore that hard
fate which reduced him to fuch neceffity,
they muft yet more ftrongly cenfure a rafh-
nefs, which could tempt him to rifk, in fo
effential a manner, the welfare of his family.

As WILSON exhibited to the public, a
fingular proof of his knowledge in this his
principal object, I am inclined to believe,
that he muft, originally, either have had
fome grammar education, or, impelled by
his genius, muft afterwards have acquired
a knowledge of the *Latin* language. How
elfe (except on the fuppofition of extraordinary affiftance, of which I have no information) could he have made ufe of MORISON's
" Hiftory," or have tranflated RAY's *Synopfis!* In 1744, he publifhed " A SY-
" NOPSIS OF BRITISH PLANTS in Mr.
" RAY's METHOD; with their Characters,
" Defcriptions, Places of Growth, Time of
" Flowering, and phyfical Virtues, accord-
" ing to the moft accurate Obfervations,
" and the beft modern Authors; together
" with a Botanical Dictionary, illuftrated
                                    " with

" with feveral Figures. By *John* WILSON.
" *Newcaftle upon Tyne.*" 8°. 1744. pp.
272,

Throughout this work, the author has
prefixed copious characters to each genus,
taken, as it appears, from RAY and TOURNE-
FORT; into many of which, in conformity
to RAY's method, he introduces the form
of the leaves, and the habit of the plant.
By this means, having added, in moft inftances, fhort defcriptions of the fpecies, his
book was an ufeful pocket manual, as far
as it extended; for he begins with the Capillary plants, and ends with the Bulbous
rooted. He fubjoins the particular places
of the rare plants in the northern parts of
England, from his own obfervations, and,
partly from a manufcript of Mr. LAW-
SON's. His remarks on the properties and
virtues, additional to thofe from RAY, he
has principally extracted from MILLER's
" *Botanicum Officinale.*"

WILSON has made fome tranfpofitions in
the diftribution of his fubjects in this volume, which prove that he had attentively
examined plants, and was well acquainted
with

with the fyftem of RAY. Some of his alterations will ftand the teft of modern accuracy, though others may be lefs happy.

He has placed all the fpecies of the *Fu-
maria* genus together, in the *Papilionaceous*
clafs; and, agreeably to the hint which
DILLENIUS gives in the *Synopfis*, p. 316,
has referred the *Plantains*, and *Sponges*, to
the *Monopetalous* flowers fucceeded by dry
feed veffels. The removal of the *Lyfima-
chiæ filiquofæ*, the two *Papavera corniculata*,
the *Chelidonium*, and the *Balfamine*, to the *Si-
liquofe* or Tetradynamous clafs of LINNÆUS,
is lefs to be approved. By thefe changes,
he has nearly annihilated RAY's twenty-
fecond clafs of *Britifh* herbs. In tranf-
pofing of fpecies, he has made more numerous alterations; fome of which are fufficiently juftified by modern improvements.
Thus he has brought under one genus the
Scordium and *Scorodonia*. He has referred
the *Raphanus rufticanus* to the *Cochlearia*
genus, as TOURNEFORT had done. The
Chelidonium genus is feparated from the
Papaver, and a new characeriftic note
framed, but the name *Papaver corniculatum*
preferved.

preferved. The only two plants met with
in this book, which do not occur in the
Synopfis of RAY, are fuch as have a doubtful title to the appellation of indigenous:
they are the *Valeriana rubra,* and *Allium
Schœnoprafum.*

I believe he died about the year 1750,
or foon after. He left the remaining part
of his work, on the *Graminaceous* and *Cryptogamous* tribes, compleat in manufcript.
In the year 1762, a perfon of *Newcaftle,*
into whofe hands the manufcript had
paffed, meditated the publication of it,
with a new edition of the work now fpoken of, which was out of print, and much
called for ; but the defign never took
effect.

CHAP. 48.

Blackftone — *His* Fafciculus Plantarum circa
Harefield — Specimen Botanicum — *Contri-*
butors to that Catalogue.

Collinfon — *a great promoter of Botany and Gar-*
dening — introduces many new productions from
America.

American *Botanifts* — Logan — Mitchell.

Warner — *His* Plantæ Woodfordienfes — *Gloffary*
to the plays of Shakefpeare — *Legacy and Exhi-*
bition to Wadham *College.*

BLACKSTONE.

IN 1737, *John* BLACKSTONE, an apothecary, in *Fleet Street, London,* publifhed " FASCICULUS PLANTARUM CIR-
CA HAREFIELD SPONTE NASCENTIUM:"
with an Appendix, containing fome fhort
notes relating to *Harefield.* 12°. pp. 118.
The order obferved in this fmall local catalogue is alphabetical, and the fynonyms
taken from *Cafpar* BAUHINE's *Pinax,* from
GERARD, PARKINSON, and others in
common ufe. Thefe are followed by the

x general

general place of growth, the particular
fpot in the inftances of rare plants, and
the time of flowering. As fcarcely any of
the *Moffes*, or *Fungi*, are introduced, the
number is fmall; only 527 fpecies. The
account of *Harefield* is very brief.

The fame author publifhed alfo, " SPE-
CIMEN BOTANICUM *quo Plantarum plurium
rariorum Angliæ indigenarum Loci natales
illuftrantur. Authore J.* BLACKSTONE."
8°. 1746. pp. 106. This fmall volume
exhibits the particular places of growth of
366 fpecies of the more rare *Englifh* plants,
and was fo far a valuable addition to RAY's
Synopfis. The arrangement is the fame as
in the *Harefield* Catalogue, and the fynonyms drawn from the fame authors; with
the addition of a few from the works of
LINNÆUS. It is embellifhed with two
elegant engravings: one reprefenting that
fingular variety of the *Clavaria Hypoxylon*,
firft figured in the *Philofophical Tranfactions*,
N° 475. and defcribed as a *Boletus:* the other
the *Lycoperdon fornicatum, Fl. Ang.* ed. 2.
p. 644; but firft defcribed and figured in
N° 474. by the late Sir *William* WATSON.
The

The *Loci natales,* or, as some modern botanists quaintly speak, the *Habitats,* of a great number of the subjects in this little work, were communicated by the friends and correspondents of the author; of whom, as they hereby contributed to enlarge the bounds of *English* botany, it is but just to record their names.

From *Yorkshire,* the author was supplied with a great number by Mr. THORNBECK, a surgeon and expert botanist, at *Ingleton,* a spot rich in the choicest objects of a curious observer. Mr. DAWSON, a surgeon of *Leeds,* communicated also many rare species : as did Mr. VERNON, of *Whitchurch,* in *Cheshire.*

The observations of the late Sir *William* WATSON, Sir *John* HILL, Dr. WILMER, and Mr. HURLOCK, contributed to enrich this little *Flora.* I find also a manuscript Catalogue of Plants growing about *Feversham* frequently referred to, written by *John* BATEMAN, A. M. This manuscript has since been the basis of a little work, published by the late *Edward* JACOB, F.S.A. under the title of " *Plantæ Favershamienses.*"
Lond.

Lond. 8°. 1777. pp. 127. To which is
annexed, a view of the Foſſil Bodies of the
iſland of *Shepey.* The plan of this catalogue
is exactly that of Mr. WARNER's, in the
" *Plantæ Woodfordienſes.*"

In this volume, Mr. BLACKSTONE has
introduced a few plants, not before recorded as natives of this iſland: ſuch are, the
*Epimedium alpinum; Ariſtolochia Clematitis;
Limonium reticulatum; Fritillaria Meleagris;*
and *Dentaria bulbifera.* Subſequent authors
have not allowed complete naturalization to
the *Epimedium,* and probably that of ſome of
the others is but of modern date. The two
laſt were obſerved by Mr. BLACKSTONE in
the environs of *Harefield.*

The author intended another volume of
the *Specimen,* for which he had collected
materials. He had alſo a taſte for Topo-
graphical Antiquities, and had made collections in that way, but did not live to
publiſh them. He died in 1753 *.

The

* Mr. HUDSON, when he ſeparated the *yellow Centory*
from the *Gentians,* gave it the name of BLACKSTONIA;
which diſtinction LINNÆUS confirmed in the *Syſtema*

The "*Specimen Botanicum*" of Mr. BLACKSTONE, I confider as the laft book publifhed in *England*, on the indigenous botany, before the fyftem of LINNÆUS had gained the afcendancy over that of RAY: nor, unlefs it were within my plan to re-count fingle papers, occafionally printed in the *Philofophical Tranfactions*, or in other collections, am I able to mention any work of importance on exotic botany, before this revolution took place, which was not built upon, or at leaft did not exhibit fome prin-ciples of, the new fyftem. Whilft this event was taking place, which cannot be computed at fewer than twenty years, com-mencing from 1740, there were, however, feveral eminent and learned men, who, al-though they did not diftinguifh themfelves by publifhing feparate tracts on the fcience, were occafionally improving it, by their

of 1767, but changed the name to *Chlora*, an appellation it had received from RENEAULME, in his *Specimen Hiftoriæ Plantarum*, publifhed in 1611. It fhould feem, that the difcovery of the true place of this plant in the fyftem, entitled Mr. HUDSON to the difpenfation of the name, or at leaft that BLACKSTONE fhould have been perpetuated in the trivial epithet.

various

various difcoveries and communications, and, ever awake to its welfare, by the patronage they extended towards it. I cannot omit to mention fome of thefe, though it be out of my power either to do fufficient juftice to their fervices myfelf, or to point out, in fome inftances, fuch memorials relating to their lives, as might properly gratify that curiofity, which efteem for their characters naturally excites.

COLLINSON.

As prior in point of time, I mention Mr. *Peter* COLLINSON, to whofe name is attached all that refpect which is due to benevolence and virtue. I have the fatisfaction of referring the reader to fome account of Mr. COLLINSON, printed in 1770 : and to further anecdotes, by Dr. LETTSOM, at the end of his " Memoirs of Dr. FOTHERGILL ;" to which is annexed, a lift of Mr. COLLINSON's papers, printed in the *Philofophical Tranfactions*, and in the *Gentleman's Magazine**. In Mr. COLLINSON's time, *England* received large acceffions to exotic botany

* See alfo a further account of Mr. COLLINSON in the *Biographia Britannica.* Vol. iv. 2d edit. p. 34.

from

from all parts of the globe; to which no one contributed more than himself, through his various correspondence, especially in *America*. He was indefatigable in his exertions to procure the seeds of curious and useful vegetables, and equally free in distributing them. Natural History in all its parts, Planting, and Horticulture, were his delight. He cultivated the choicest exotics, and the rarest *English* plants. His garden contained, at one time, a more complete assortment of the *Orchis* genus, than, perhaps, had ever been seen in one collection before. He died August 11, 1768, in the 75th year of his age *.

Numerous were the channels by which *England* was enriched with the seeds and specimens of *American* productions. BARTRAM was constantly employed in collecting. Governor COLDEN, of *New York*, and Dr. MICHELL, in *Virginia*, were frequent in their communications to MILLER, to CATESBY, to COLLINSON, and others. For

---

* The name of Mr. COLLINSON is perpetuated in a beautiful *American* plant of the *Diandrous* class, well known in the *English* gardens.

Dr.

Dr. FOTHERGILL's inceſſant exertions in the ſame deſigns, being at a later period, are too well known to be repeated here. Governor COLDEN ſent to LINNÆUS upwards of 200 ſpecies, the account of which was printed in the *Upſal Acts* for 1743 and 1744; and LINNÆUS, in his *Flora Zeylanica*, gave to a plant of the *Tetrandrous* claſs, the name of his correſpondent.

LOGAN.

Several ingenious gentlemen in *America* purſued botanical inveſtigations with great ſucceſs about this period. *James* LOGAN, Eſq; afterwards Preſident of the Council, and Chief Juſtice of *Penſylvania*, inſtituted a ſet of Experiments on the Maiz, relating to the ſexes of plants. They were firſt communicated in a letter to *Peter* COLLINSON, F.R.S. in 1735; and were printed in the *Philoſophical Tranſactions*, Vol. xxxvi. p. 192. They were afterwards enlarged, and publiſhed in *Latin*, at *Leyden*, in 1739, under the title of " *Experimenta et Meletemata de Plantarum Generatione*;" and republiſhed with an *Engliſh* tranſlation, if I miſ-

take

take not, by Dr. FOTHERGILL, in 8°. 1747.
pp. 39. They have been confidered, and
appealed to, as among the moft decifive in
eftablifhing the doctrine they were intended
to illuftrate and confirm.

M I T C H E L L.

Dr. *John* MITCHELL, then refident at
Urbana, in *Virginia,* fent over, in 1741,
the defcriptions of thirty genera of plants,
of which fix were entirely new; others were
corrected and amended. Among the moft
remarkable are, the Ginfeng of *America,*
Panax quinquefolium : the *Liquid Ambar*
Styraciflua : the *Malacodendron,* afterwards
called by CATESBY, STEWARTIA, in honour of the *Earl of* BUTE : the *Zizania*
aquatica. In the introduction, Dr. MIT-
CHELL difcourfes on the principles of botany, and appears to have paid attention to
the *Hybrid* productions. This paper was
feparately publifhed, in 4°. at *Nurenburgh,*
in 1769.

In 1743, he fent over to Mr. COLLIN-
SON, an ingenious " Effay on the Caufes of
" the different Colours of People in diffe-
                                    " rent

" rent Climates." It was defigned as a fo-
lution of the prize problem from the Academy of *Bourdeaux*; but was publifhed in
the *Philofophical Tranfactions*, Vol. xliii.
pp. 102—150.

The queftion concerning the caufe of the
black colour of the fkin in Negroes, has
exercifed the pens of many philofophers and
anatomifts. What has perplexed the queftion the more is, that thefe ingenious writers (among whom are principally *Malpighi,
Boyle, Winflow, Meckel,* and *Barrere)* have
differed about matters of fact that fhould
feem to be cognizable by the fenfes.

It would be improper in this work to
purfue the learned author through all his
ingenious details and curious *fcholia* on this
fubject; it muft be fufficient to obferve,
that, on the *Newtonian* doctrine of the caufes
of colours, he deduces the colour of the
fkin of Negroes from the ftructure, after
eftablifhing certain propofitions: 1. That
the colour of White People proceeds from
the colour which the *epidermis* tranfmits.
2. That the denfity of the fkins of Negroes
allows of no tranfmiffion of colour. 3. The

T 4   part

part of the fkin which appears black in Negroes, is the *corpus reticulare cutis*, and external *lamella* of the *epidermis*. 4. That the colour does not proceed from any black humour or fluid parts contained in their fkins. 5. That the *epidermis*, efpecially its external *lamella*, is divided into two parts, by its pores and fcales, 200 times lefs than the particles of bodies, on which their colours depend. Hence Dr. MITCHELL concludes, " that the proximate caufe of the colour of Negroes is threefold; *viz.* the opacity of their fkins, proceeding from the thicknefs and denfity of the texture, which obftructs the tranfmiffion of the rays of light from the white and red parts below them; together with their greater refractive power, which abforbs thofe rays; and the fmallnefs of the particles of their fkins, which hinder them to reflect any light." After which, he difcourfes on the influence of the fun, and the modes of life among the inhabitants of hot countries, as the remote caufes of the colour of Negroes and Indians.

Dr. MITCHELL returned to *England*, I
believe,

believe, about the year 1747 or 1748; became a Fellow of the Royal Society; and
was the writer of an inftructive memoir "On
" the Preparation and Ufes of the various
" Kinds of Pot-Afh." *Phil. Tranf.* Vol.
xlv. p. 541—563. And of " A Letter
" concerning the Force of Electrical Co-
" hefion." Vol. li. p. 390.

WARNER.

Richard WARNER, Efq; of *Woodford-
Row,* in *Effex,* merits a particular remembrance at this period, for his regard to the
fcience of botany, and the refpect and honour he ever fhewed to the lovers of it.
" He was bred to the law," as we are informed in the ' Anecdotes of Mr. *William*
BOWYER,' " and had chambers in *Lin-
" coln's Inn*; but, being poffeffed of a genteel
" fortune, refided at a good old houfe on
" *Woodford Green.*" Here he maintained a
botanical garden, and was very fuccefsful in
the cultivation of rare exotics. He was
not unacquainted with indigenous plants.
The herborizations of the Company of
Apothecaries were, once in the feafon,
ufually

ufually directed to the environs of *Wood-ford*, where, after the refearches of the day, at the table of Mr. WARNER, the products of *Flora* were difplayed. The refult of the inveftigations made in that neighbourhood, was publifhed by Mr. WARNER, under the title of " *Plantæ Wood-* " *fordienfes*; or, a Catalogue of the more " perfect Plants growing fpontaneoufly " about *Woodford*, in *Effex.*" *Lond.* 1771. 8°. pp. 238. As none of the *Graminaceous*, or *Cryptogamous* tribes, are introduced, the lift does not exceed 518 fpecies. The order is alphabetical, by the names from RAY's *Synopfis*; after which follow the fpecific character at length, from HUDSON's " *Flora Anglica*," the *Linnæan* clafs and order, the *Englifh* name, place, and time of flowering. In the Preface, the author enumerates the names of more than twenty of his friends, among whom are many of thofe alluded to above, by whofe joint affiftance he was enabled to enlarge his work beyond what his own obfervations might otherwife have allowed. Mr. WARNER was alfo diftinguifhed for his polite learning;

learning; and eminently fo, for his critical knowledge in the writings of *Shakefpeare*, of whofe plays he had long meditated to give a new edition; but defifted, on the appearance of Mr. *Steevens*'s propofals. In 1768, he publifhed " A Letter to *David* " *Garrick*, Efq; concerning a Gloffary to the " Plays of *Shakefpeare*." 8°. This Gloffary he continued to augment, to the laft days of his life. He tranflated the Comedies of *Plautus*, left undone by *Thornton*, which were publifhed in 1772 and 1774.

Mr. WARNER, in his youth, as is related of the great LINNÆUS, had been remarkably fond of dancing; nor, till his paffion " for that diverfion fubfided, did he " convert the largeft room in his houfe " into a library." He died April 11, 1775; and bequeathed his valuable books to *Wadham* College, *Oxford*, where he received his education; and left to the fame Society an exhibition for a botanical lecture.

CHAP. 49.

Ehret—*a* German *of the marquifate of* Baden
Durlach —*firſt patronized by* Trew — *Paints*
plants in the Royal Garden of Paris — *and in*
Clifford'*s garden under* Linnæus — *Settles in*
England—*Patronized by the* Literati—Plantæ
Selectæ *of* Trew *painted by him —*Ehret'*s pub-*
lications — *His papers in the* Philoſophical
Tranſactions.
Hill—*his writings.*

E H R E T.

AMONG the various contingencies
which favoured the introduction of
the *Linnæan* ſyſtem into *England,* it is not
unimportant to mention the effect of the
admirable pencil of the late Mr. EHRET.
This ingenious artiſt brought with him,
not only a general taſte for botany, but a
particular knowledge of the principles, on
which the ſyſtem of LINNÆUS was founded; and was among the firſt who diſplayed
it, in the ſpecimens of his art.

The father of *George Dyoniſius* EHRET
was gardener to the Prince of *Baden Dur-*
lach.

lach*. Young EHRET very early shewed a
taste for drawing, and painting the flowers
of the garden. And although he received
no instructions, yet such was his proficiency, that, whilst a very young man, he had
painted 500 plants with a skill and accuracy that was almost unexampled, under
the disadvantages of so total a want of in-
struction as our young artist had experienced. His merit, however, remained long
unknown, or at least ineffectually noticed,
until it was discovered by a gentleman of
curiosity and judgment, who visited the
garden, of which his father was the superintendant. Fortunately for young EHRET,
this stranger was a physician and a friend
of the celebrated Dr. TREW, of *Norimberg*,
to whom he justly supposed these paintings
would be acceptable. EHRET by this means
was introduced to TREW, who immediately
purchased the whole 500 paintings, and generously gave him double the price at which
the young artist had modestly valued them.

---

* *Charles*, Prince of *Baden Durlach*, was a patron of
botany, and his garden was famous at that time. He
sent his principal superintendant of the garden, on the unfortunate expedition with HEBENSTREIT, into *Africa*.

The

The liberality of TREW, by which EHRET put 4000 florins into his pocket, infpired him with confidence in his own abilities, and fuch a fhare of ambition as inclined him to quit his home, and feek at once to raife his fortune, and to gratify the defire he had to fee the world. It appears that he was too much elated with his fuccefs; and, as the effect of fome fhare of vanity, and a want of œconomy not unufual in young men, he foon diffipated this fum, and, in queft of adventures, went to *Bafil,* with the laft, and thofe only a few, of his florins in his pocket. Here, fhutting himfelf up, he, with great diligence, and fingular exertion, ftimulated now by preffing neceffity, foon exhibited numerous fpecimens of his art; and, though he had learned to fet a higher value upon them, found a demand beyond his induf-try to fupply. Having thus recruited his finances, he journeyed into *France,* and re-fided fome time at *Montpelier,* where he taught his art to a lady of fortune, who re-warded him generoufly, and, on his wifh to remove, paid his expences to *Lyons* and *Paris.* At the latter city he became known to JUSSIEU, and was for fome time employed

to

to paint the plants of the Royal Garden, under that eminent Profeſſor's infpection. After a certain time, he exchanged his fitu ation at *Paris*, for that of *London*; but not fucceeding to his mind, he foon returned to the continent. The precife time of his being firſt in *England*, I cannot afcertain; but it was, I conjecture, before his employment in the garden of Mr. CLIFFORD, where LINNÆUS found him in the year 1736. From LINNÆUS himſelf he was taught attention to the parts of the flower, and hence became early inſtructed in the principles of the fexual fyſtem. His fine tafte, and botanical accuracy, were, I ap- prehend, firſt publicly manifefted in the figures of the *Hortus Cliffortianus*, pub- liſhed in 1737; and, from that time, EH- RET became ſtrongly attached to the prin- ciples of the *Swede*.

He returned to *England* about the year 1740, or foon after that period: and here he fpent the remainder of his days. His firſt patron in this country was *Taylor* WHITE, Efq; for whom he finiſhed 300 paintings of plants. He foon after procured the patro- nage of Dr. MEAD, for whom he painted

200, and who generously advanced his price. In consequence of this countenance and protection, he obtained encouragement from Sir *Hans* SLOANE, and many other opulent lovers of his art.   Dr. FOTHERGILL procured large collections from him; and the late eminent patroness of natural history, the *Duchess of* PORTLAND, possessed, besides near 300 paintings of exotics, upwards of 500 of *English* plants, done on vellum, and highly finished, by this admirable artist.

Another of his patrons, and to whose obliging information I owe great part of the foregoing anecdotes relating to him, was *Ralph* WILLETT, Esq; of *Merly*, in *Dorset-shire*; at whose seat Mr. EHRET was accustomed, for many years, to spend several weeks in the summer season, and in whose friendship Mr. EHRET reposed, as executor in the last arrangement of his affairs.   The library at *Merly* exhibits a copious collection of exotics, done by EHRET: not fewer than 230 finished specimens on vellum; besides seventy on paper; and more than 500 in an unfinished state.

The first published specimens of his pencil, after his settlement in *England*, that I

am

am acquainted with, were exhibited in the
44th volume of the *Philofophical Tranfactions*,
N° 478. for January and February 1746;
by the figure of the *Keratophyton flabelli-
forme* of RAY (Gorgonia verrucofa *Lin.*)
for a paper written by Sir *Hans* SLOANE :
and by two excellent figures of the *Oenanthe crocata,* and *Cicuta virofa,* in the
fame volume, intended to illuftrate Mr.
WATSON's obfervations on the fatal Qualities of thofe Plants.

Very early after his arrival in this kingdom, he began to paint figures of the rareft
products of the *Englifh* gardens, for his friend
and firft patron Dr. TREW; for whom, in
the end, he finifhed 300. Of thefe, at
different periods, 100 were engraved, and
publifhed in *Decads,* under the following
title:

" PLANTÆ SELECTÆ, *quarum Ima-
gines ad exemplaria naturalia Londini in
hortis curioforum nutrita, manu artificiofa
pinxit Georgius Dionyfius* EHRET, *Germa-
nus, collegit nominibus notifque illuftravit
Chr. Jacob.* TREW, *M.D. Norib. in Æs in-
cidit et vivis coloribus reprefentavit Jo. Jac.*

Haid. Augustanus." Decuria I. 1750. fol.
reg.—Decur. X. 1773.

Seven *Decads* of this work were published at Dr. TREW's expence, during his life-
time; and the remaining three by Dr. VO-
GEL, after his decease. The whole is executed in so splendid a manner, as to consti-
tute, at this day, one of the finest ornaments of the botanical library.

The only publication of any importance
in *England,* in which Mr. EHRET was engaged throughout, as far as I can find, was
BROWN s " Natural History of *Jamaica,"*
printed in 1756, for which he drew all the
figures, amounting to 40 tables. As they
were principally taken from prepared and
dried specimens, they cannot be numbered
among his capital performances.

Mr. EHRET drew, and himself engraved,
a set of tables of Exotics, two or three on
each plate, to the number of fifteen; each
table containing also a Butterfly of exotic
origin. These were published at *London,*
in 1748—1759. The last of these exhibits the Cape Jasmine, *Gardenia florida,*
which had flowered for the first time in
England,

England, in the garden of Mr. WARNER, at *Woodford,* in the year 1758. A defcription of this elegant plant; the generical character of the *Laurus Saffafras*; and the defcription of a new *Lithofpermum,* all written by Mr. EHRET, were printed in the "*Nova Aƈta Academiæ Curioforum.*" Tom. II. *Norimb.* 1761.

An Account of the *Ophrys fcapo nudo foliis radicalibus ovato-oblongis, dimidii fcapi longitudine,* defcribed by GRONOVIUS in his "*Flora Virginica*;" with a figure. Vol. liii. p. 81. The *Ophrys lilifolia* of LINNÆUS: it was fent from *Philadelphia* by Mr. BARTRAM, and flowered in *England,* for the firft time, in the garden of Mr. COLLINSON, in the year 1758.

An Account of a new *Peruvian* plant lately introduced into the *Englifh* gardens; with a figure. Vol. liii. p. 131. This is the *Nolana proftrata* Lin. which flowered in the garden at *Chelfea,* for the firft time in *England,* in 1761, now very common.

A Defcription of the *Andrachne,* with its botanical character, and a figure. Vol.

lvii. p. 114. The *Arbutus Andrachne,*
which firſt flowered in *England,* in 1766,
in the garden of Dr. FOTHERGILL.

His ingenuity and knowledge of nature
raiſed him to a degree of reputation among
the literati, and obtained him the diſtinction
of being choſen a Fellow of the Royal Society. Beſides the profit accruing from
thoſe numerous exhibitions of his pencil, he
applied for many years, with great aſſiduity, to the buſineſs of teaching his art ; and
if his ingenuity did not meet with a reward
equal to his merit, yet his labours, in the
end, proved ſufficiently lucrative, to afford
him a moderate independence ; though, to
the laſt, he ceaſed not to employ his pencil.

He died in September 1770, in the 60th
year of his age *.

Mr. EHRET married the ſiſter of *Philip
Miller,* of *Chelſea,* by whom he left one
ſon.

* Mr. EHRET was complimented by Dr. TREW, in
the Third Decad of the *Plantæ Selectæ,* with a new genus, which he called by his name. The EHRETIÆ are
trees of the *Pentandrous* claſs, firſt deſcribed and figured
by SLOANE ; to which, new ſpecies have been added by
JACQUIN.

He

He was well verfed in the botany of this
country, and delighted in painting the indigenous plants. He was ever beft pleafed
when employed by fcientific people ; fince
his wifh was always to follow nature, and
to exhibit on his piece the true characters,
without the fmalleft deviation for the fake
of embellifhment. Having early imbibed
the principles of *Linnæus's* fyftem, he attended to the difcrimination of the parts on
which it was founded, with an accuracy
that commanded obfervance; and while his
excellence in delineating and painting drew
admiration, and diffufed a tafte for the
ftudy of plants, the truth of his pencil in-
ftructed thofe who beheld it in the principles of the fcience.

HILL.

About the year 1751, Dr. HILL began
to publifh on the fubject of botany. His
" Hiftory of Plants," printed in that year,
although compiled and tranflated principally from LINNÆUS, was not adapted to
indigenous botany, nor fufficiently calculated to inftruct the ftudent in the ultimate

U 3   part

part of any fyftem, the fpecific diftinctions;
fince LINNÆUS had not as yet completed
the exemplification by modelling the character throughout the whole ; the *Species
Plantarum* not being publifhed till the year
1753.

I mean not to enter on any detail of his
numerous writings, fince they are well
known, and moft of them pofterior to the
limits of my plan. Although it may be
difficult to reconcile the praifes this author beftows on LINNÆUS, in many of his
writings, with the cenfures contained in his
" Britifh Herbal," yet his works had a favourable influence in promoting the fcience
in general, though not the *Linnæan* modifi-
cation of it in particular *.

* For an account of Sir *John* HILL, I refer the reader
to the *Biographica Dramatica*. Edit. the 2d. 1782.

CHAP. 50.

Sir William Watſon — *Anecdotes of — His early bias to Natural Hiſtory — Admitted into the* Royal Society—*Diſtinguiſhes himſelf as a Botaniſt—His papers on that ſubject in the* Philoſophical Tranſactions — *Publiſhes* Peyſſonnel's Diſcoveries on Zoophytes — *Appointed one of the Truſtees to the* Britiſh Muſeum *by* Sloane *himſelf — One of the Revivers of Electricity — Makes ſeveral eminent diſcoveries in that branch of philoſophy—His papers on that ſubject printed in the* Philoſophical Tranſactions.

WATSON.

AMONG thoſe learned botaniſts of *England,* who early recognized the prevailing excellencies of the *Linnæan* ſyſtem, muſt be ranked the late Sir *William* WATSON. At a period when Botany was feebly ſupported in theſe kingdoms, after the deceaſe of the SHERARDS, and the retirement of SLOANE, his talents and his zeal enabled him, as far as the influence of

U 4

an

an individual could extend, to fuftain and
promote this fcience, not only with his
own countrymen, but with thofe learned foreigners who vifited this kingdom.
Whilft, therefore, juftice to his character
and attainments, in the fubject of this
work, demand confideration, I feel an additional motive to pay a tribute to his memory, arifing from a grateful remembrance
of the friendfhip and correfpondence with
which he honoured me.

Sir *William* WATSON was born in 1715,
in *St. John's Street*, near *Smithfield*. His father was a reputable tradefman in that ftreet,
and died, leaving him very young. When
he had attained to a proper age, he was
fent to Merchant Taylor's School ; and from
thence was apprenticed to Mr. *Richardfon*,
apothecary, in 1730.

In his youth he had a ftrong propenfity
to the ftudy of natural hiftory, and particularly to that of plants. This led him to
make frequent excurfions in a morning, feveral miles from *London*; fo that he became
early well acquainted with the *Loci natales*
of the indigenous plants of the environs of
London ;

London ; and, during his apprenticeſhip, he gained the honorary premium given annu- ally by the Apothecaries Company, to ſuch young men as exhibit a ſuperiority in the knowledge of plants, in thoſe excurſions made by the Demonſtrator of *Chelſea* Gar- den; and inſtituted for the purpoſe of ini- tiating the apprentices of the Company in a ſcience ſo neceſſary to the profeſſion. This premium, as hath been obſerved in the courſe of the preceding pages, conſiſted of a handſomely bound copy of RAY's *Synopſis.* He continued, at times, throughout his life, to attend on theſe occaſions, and meet his former aſſociates with great pleaſure and delight.

In 1738, Mr. WATSON married, and ſet up in buſineſs for himſelf. His ſkill, his activity, and diligence in his profeſſion, ſoon diſtinguiſhed him among his acquaintance; as did his taſte for natural hiſtory, and his general knowledge of philoſophical ſubjects among the members of the Royal Society, of which honourable body he was elected a member early in the year 1741 ; his two firſt communications being printed in the

4 1ſt

41ſt volume of the *Philoſophical Tranſactions.*

Soon after his admiſſion into the Royal Society, Mr. WATSON diſtinguiſhed himſelf as a botaniſt. His earlieſt paper on this ſubjeƈt was, " An Account of the ce-
" lebrated HALLER's *Enumeratio Stirpium*
" *Helvetiæ,* extraƈted from the *Latin,* and
" illuſtrated with a *Conſpeƈtus* of the au-
" thor's method, and with various obſer-
" vations." This was printed in the *Phi-loſophical Tranſaƈtions* (*a*).

In the ſame volume (*b*), and in the ſuc-ceeding (*c*), he excited the attention of the curious in this way, by ſome " Critical Re-
" marks on the Rev. Mr. PICKERING's
" Paper concerning the Seeds of Muſh-
" rooms," which, that gentleman having ſeen a ſhort time before, conſidered as a new diſcovery; whereas Mr. WATSON ſhewed, that they had been demonſtrated ſeveral years prior to that period, by M. MICHELI, in his " *Nova Plantarum Genera,*" printed at *Florence* in 1729.

(*a*) Vol. xlii. p. 369—80.   (*b*) p. 599.
(*c*) Vol. xliii. Nº 473. p. 51.

But

But that which attracted the attention of foreign botanists particularly, was his description of a rare and elegant species of *Fungus*, called from its form *Geaster* (*d*). This was written in *Latin*, and accompanied with an engraving. It has since been called *Lycoperdon fornicatum.*

In the same volume are inserted some very instructive observations on the *Cicuta*, or Common Hemlock; occasioned by the death of two of the *Dutch* soldiers; quartered at *Waltham Abbey*, in *Essex*; which happened in consequence of their having eaten this herb instead of Greens (*e*).

The death of two of the *French* prisoners in 1746, occasioned by their eating the roots of the *Hemlock Dropwort*, produced from Mr. WATSON a paper, which in an eminent manner exemplified his skill in the knowledge of plants. It abounds with curious and critical observations on that plant, and on the *Sium Erucæ folio* of *Caspar* BAUHINE (Cicuta virosa *Lin.*) with which it had been frequently confounded;

(*d*) *Phil. Transf.* Vol. xliii. p. 234. t 2. f. 11.
(*e*) Ibid. N° 473. p. 18.

as both had also been commonly mistaken for *Water Parsnep*. It is accompanied with engravings of the plants, from the excellent drawings of Mr. EHRET (*f*).

In the 45th volume of the *Philosophical Transactions*, is printed a Translation, by Mr. WATSON, of a Letter to Sir *Hans* SLOANE, from Dr. GARCIN, of *Neuchatel*, containing a complete history of the *Cypress* of the antients ; the *Henna*, or *Alcanna* of the *Arabians*, called by LINNÆUS *Lawsonia inermis*; a Shrub, famous for its use, both in medicine, and as a dye, all over the East, insomuch that, at *Constantinople*, the duty on this drug amounts to 18,000.ducats annually (*g*).

In 1748, Mr. WATSON had an opportunity of shewing attention to M. KALM, during his abode in *England*, which was from February till August, when he embarked for *America*. He introduced him to the curious gardens, and accompanied him in several botanical excursions in the environs of *London*. This eminent pupil of LINNÆUS,

(*f*) *Phil. Transf.* Vol. xliv. p. 227—245.
(*g*) Ib. Vol. xlv. p. 564—578.

who

who was a *Swedifh* divine, on his return home, became Profeffor of Oeconomy at *Abo*, where he died Nov. 16, 1779, aged 63.

The fame civilities were manifefted by Dr. WATSON to the prefent eminent Dr. PALLAS, of *Peterfburgh*, during his abode in *England*, which was from July 1761, to April 1762.

In 1749, in company with Dr. MIT-CHELL, Mr. WATSON examined the remains of the garden, formerly belonging to the TRADESCANTS; of whom, fee chap. 14. of this work. They found the *Arbutus*, and the *Cupreffus Americana*, with other exotics, in a vigorous ftate, after having fuftained the winters of this climate for 120 years. This fituation had alfo afforded a proof, not often exemplified, of the large fize to which the *Common Buck-Thorn* will grow. They found one about 20 feet high, and near a foot in diameter (*h*).

In 1751, were laid before the public, fome very curious and interefting particulars, relating to the fexes of plants, which

(*h*) *Phil. Tranf.* Vol. xlvi. p. 160.

tended

tended to confirm the truth of that doctrine in a remarkable manner. Thefe were occafioned by a letter from Mr. MYLINS, of *Berlin*, informing Mr. WATSON, that a tree of the *Palma major foliis flabelliformibus*, which, although it had borne fruit for 30 years paft, had never brought any to perfection, until the flowers of a male-tree, brought from *Leipfic*, 20 *German* miles diftant, had been fufpended over its branches. After this operation, the tree yielded, the firft year, above 100, and the fecond, upon repeating the experiment, above 2000 ripe fruit; from which 11 young Palm-trees had been propagated (*i*).

In the fame volume are fome remarks on the cafe of two women in *Brabant*, who had been nearly poifoned by eating the leaves of what had been called *White Henbane*; but Mr. WATSON proved, that it muft have been the *Hyofcyamas niger*, fince the *white* does not grow fpontaneoufly in that country. The fame letter confirms the poifonous effect of the *Yew-tree* upon horfes (*k*).

(*i*) *Phil. Tranf.* Vol. xlvii. p. 169.
(*k*) Ib. Vol. xlvii. p. 199.

Mr.

Mr. WATSON paid the fame tribute, in 1751, to the memory of Dr. *Henry* COMP-TON, Bifhop of *London*, the friend and patron of Mr. RAY, as he had done to that of the TRADESCANTS; and gives a lift of 33 exotic trees, which were then remaining in the garden at *Fulham.* From this catalogue may be inferred, not only the original fplendour of the garden, and the zeal and tafte the Bifhop fhewed in the cultivation of fuch numerous curiofities, but the facility with which trees of very different latitudes may become naturalized in *England (l).*

In the fame volume, page 301, we find " An Account of the Cinnamon Tree;" .occafioned by a large fpecimen, equal in fize to a walking cane, fent over by Mr. ROBINS to Dr. LETHERLAND, and which was exhibited to the infpection of the Royal Society. From this Account we learn, that three Cinnamon Trees, which were intended to have been fent to *Jamaica,* were growing in the garden of *Hampton Court* in the reign of King *William.*

(*l*) *Phil. Tranf.* Vol. xlvii. p. 241—247.

In

In the year 1752, Mr. WATSON laid
before the Royal Society two rare *Englijh*
plants ; the *Lathræa Squamaria,* and the
Dentaria bulbifera : the latter unnoticed
both by Mr. RAY and DILLENIUS. Thefe
were difcovered by Mr. BLACKSTONE,
near *Harefield* (*m*).

He alfo defcribes, in this volume, that
fingular vegetable produ&ion, noticed be-
fore under the article of *Thomas* KNOWL-
TON, as firft difcovered by him, and called
Moor Balls, the *Conferva Ægagropila* of
LINNÆUS (*n*).

Mr. WATSON, about this time, was the
firft, as I apprehend, who communicated to
the *Englijh* reader, an Account of a Revolution which was about to take place
among the learned, in Botany and Zoology, refpe&ing the removal of a large body
of marine produ&ions, which had heretofore been ranked among vegetables ; but
which were now proved to be of animal
origin, and ftand under the name of Zoophytes, in the prefent *Syftem of Nature.* It

(*m*) *Phil. Tranf.* Vol. xlvii. p. 428.
(*n*) Ib. p. 498.

may

may be eafily feen that this refpects the
Corals, Corallines, *Efcharæ*, Madrepores,
Sponges, &c. and although even GESNER,
IMPERATUS, and RUMPHIUS, had fome
obfcure ideas relating to the dubious ftructure of this clafs, yet the full difcovery, that
thefe fubftances were the fabrications of
Polypes, was owing to M. PEYSSONNEL,
phyfician at *Guadaloupe*. This gentleman
had imbibed this opinion firft, in 1723, at
Marfeilles; and confirmed it, in 1725, on
the coaft of *Barbary.* While in *Guadaloupe*, he wrote a volume of 400 pages in
4°. in proof of this fubject, which he tranfmitted in manufcript to the Royal Society of *London.* This treatife, in which
the author feemed to have put the matter
out of doubt, as to the animal origin of
thefe bodies, was tranflated, analyzed, and
abridged, in 1752, by Mr. WATSON;
and publifhed in the *Philofophical Tranfactions* (*o*), at a time when the learned
were wavering in their opinions on this
matter. M. TREMBLY's inveftigation re-

(*o*) Vol. xlvii. p. 445—469.

fpecting the Frefh Water Polypes had pav-
ed the way for the reception of PEYSSON-
NEL'S truths; and Mr. WATSON himfelf,
in company with M. TREMBLY, had an
opportunity, on the coaft of *Suffex*, when
on a vifit at the Duke of RICHMOND'S, in
one of thofe annual excurfions (*p*) which
for many years he feldom failed to make in
the fummer feafon, of verifying M. PEYS-

(*p*) It may gratify the curiofity of fome, who reverence
the name of Mr. RAY, to be informed, that in one of
thefe excurfions, Dr. WATSON was led, by his refpect to
the memory of that great and good man, to vifit the fpot
where he had lived at *Black Notley*, in *Effex*. This was
in the year 1760. To Dr. WATSON this was claffical
ground. I was informed by him, at that time, that he
found Mr. RAY's monument removed out of the church,
where it formerly ftood, into the church-yard, and hardly
vifible for brambles: thefe he had removed while he
ftayed. That he found the houfe in a ftate which indicated
no alteration having taken place, except what more than
half a century of time might be fuppofed neceffarily to
have occafioned; unlefs that indeed fome of the windows
were ftopped up to fave the tax; and that the orchard
bore all the appearance of being, as near as poffible, in the
ftate in which it muft have been in Mr. RAY's life-time.
That the inhabitants of the village knew little of him;
and the people of the houfe had only heard that he was a
great traveller.

SONNEL'S

sonnel's fyftem, in viewing the Polypes of the Corallines.

In 1753, was printed, " An Account of " the Second Volume of the *Flora Sibirica* " of Gmelin ;" exhibiting fome extracts relating to the cure of the venereal difeafe, in *Siberia*, by the decoction of a fpecies of *Cirfium*, and an *Iris :* and on the diftillation of a fpirituous liquor from the (*q*) *Spondylium*, or Cow-Parfnep.

In the fame volume of the *Tranfactions* (*r*), fome Obfervations, tending to determine what was the *Byffus* of the antients ; occafioned by a fubftance which was fent over by Profeffor Bose, of *Wittemberg*. It proved to be no other than the common *Byffus velutina*, in a bleached ftate ; whereas the *Byffus* of the antients was judged by Mr. Watson to be, moft probably, a Cotton ; which is confirmed in a very elaborate and critical Differtation, written by Dr. *Reinhold* Forster, and publifhed in 1776.

Remarks, additional to thofe of Dr. Martyn, on the Sex of the *Holly-Tree* ;

(*q*) *Phil. Tranf.* Vol. xlviii. p. 141—152.
(*r*) Ib. p. 358.

X 2   which

which juftified the removal of it from the *Tetrandrous* to the *Potygamous* clafs (*s*).

"Some Obfervations upon the *Agaric*
" lately applied after Amputations, with re-
" gard to the determining its Species (*t*)."
Some doubts had arifen relating to the exact
fpecies of the *Styptic Agaric*, which had
juft then excited the attention of the furgeons, both in *France* and *England*. Mr.
WATSON having written afterwards to
M. Bernard de JUSSIEU at *Paris*, was affured that the *French* furgeons had ufed the
Agaricus pedis equini facie of TOURNEFORT,
which is the *Boletus igniarius* Lin. (*u*).

In 1754, Mr. WATSON wrote an Account of the firft Edition of the *Species
Plantarum* of LINNÆUS; which was publifhed in the *Gentleman's Magazine*, p. 555
for that year. It is not only highly worthy
of being read, for the ufeful information,
and curious critical matter it contains; but
alfo on account of its having produced from
that celebrated Profeffor, a handfome letter,

(*s*) *Phil. Tranf.* Vol. xlviii, p. 615.
(*t*) Ib. p. 811.
(*u*) Ib. Vol. xlix, p. 23.

written

written in *Latin*; in which he takes occafion to acknowledge the candour, and fkill
of the author, in high terms; and vindi-
cates himfelf for having, in his work above-
mentioned, given to the *Meadia* (a plant fo
called by CATESBY, in honour of Dr.
MEAD) a different name. LINNÆUS's
Letter was printed the fucceeding year, in
the fame publication (*w*).

In 1758, he had occafion to confirm the
fatal effects of the *Oenanthe crocata*, or
Hemlock Dropwort, by the death of a perfon at *Havant*, in *Hampfhire*, from having
taken about four fpoonfuls of the juice of
the root, inftead of that of the Water-Parf-
nep. It was obferved, that in this inftance, as in that of the *French* prifoners,
all the fufferers were affected with the
locked jaw (*x*).

Thefe talents, it may be eafily imagined,
rendered him a welcome vifitor to Sir
Hans

(*w*) *Gent. Mag.* Vol. xxv. p. 317.
(*x*) *Phil. Tranf.* Vol. l. p. 856—9.
I take this opportunity to remark alfo, that, in the cafe
of a young woman poifoned by the fame means, which
is printed in the 5th volume of the *London Medical Jour-*

nal,

Hans SLOANE, who had retired to *Chelsea* in 1740. In fact, he enjoyed no small share of the favour and esteem of that veteran in science; and was honoured so far, as to be nominated one of the Trustees of the British Museum by Sir *Hans* himself, who died Jan. 11, 1753.

After its establishment in *Montagu House*, Mr. WATSON was very assiduous, not only in the internal arrangement of subjects, but also in getting the garden furnished with plants; insomuch that, in the first year of its establishment, in 1756, it contained no fewer than 600 species, all in a flourishing state.

Having given ample specimens of the genius and abilities of Mr. WATSON, as a naturalist, we must now consider his talents in some other branches of knowledge. Among these, nothing contributed so much to extend his fame, and enlarge his connexions with men of science, as his discoveries in

---

nal, p. 192—193. subsequent enquiry has convinced me, that the incapacity of swallowing, with which she was affected before her death, arose from the same affection of the jaw.

electricity.

electricity. He became early enamoured with the phænomena of this wonderful agent in nature; an attention to which had been fome time before excited, among the philofophers of *Europe*; and particularly in *England*, by Mr. *Stephen* GRAY, of the Charter-Houfe; *Granville* WHELER, Efq; Dr. DESAGULIERS; and others.

About the year 1744, Mr. WATSON took it up, and made feveral important dif-coveries in it. At this time, it was no fmall advancement in the progrefs of elec-tricity, to be able to fire fpirit of wine. He was the firft in *England* who effected this, and he performed it, both by the *direct*, and the *repulfive* power of electricity. He afterwards fired inflammable air, gunpow-der, and inflammable oils, by the fame means. He alfo inftituted feveral other experiments, which helped to enlarge the power of the electrician; but the moft important of his difcoveries was, the prov-ing, that the electric power was not created by the globe or tube, but only collected by them. Dr. FRANKLIN, and Mr. WIL-

SON, were alike fortunate, about the same time. It is easy to see the extreme utility of this discovery in conducting all subsequent experiments. It soon led to what he called " the circulation of the electric " matter."

Besides these valuable discoveries, the Historian of Electricity informs us, that Mr. WATSON first observed the different colour of the spark, as drawn from different bodies; that electricity suffered no refraction in passing through glass; that the power of electricity was not affected by the presence or absence of fire, since the sparks were equally strong from a freezing mixture, as from red hot iron; that flame and smoke were conductors of electricity; and that the stroke was, as the points of contact of the non-electrics on the outside of the glass. This investigation led to the coating of phials, in order to increase the power of accumulation; and qualified him eminently to be the principal actor in those famous experiments, which were made on the *Thames*, and at *Shooter's Hill*, in the

years

years 1747 and 1748; in one of which, the
electrical circuit was extended four miles,
in order to prove the velocity of electricity;
the refult of which convinced the atten-
dants that it was inftantaneous (*y*.)

It ought alfo to be remembered, that

(*y*) ' Thefe, and other experiments, were made in fo
' great a ftyle, and with fuch fuccefs, as to draw the ap-
' probation and applaufe of almoft all fucceeding philo-
' fophers in that branch.   Among others, the celebrated
' VOLTA has given him teftimony of the excellence and
' greatnefs of his experiments, in a paper publifhed within
' thefe few years.   In that paper, he fhews how fimple
' electrical conductors might be fo conftructed, as not
' only to give fhocks like the *Leyden* phial, but even fuch
' as are fufficiently powerful to kill large animals, and to
' equal the effects of lightning.   He however expreffes
' his defpair of ever feeing fuch put into execution ; but
' adds—" *Un* WATSON *forfe fat ebbe tentato di farlo,* &c.
" A WATSON perhaps might be tempted to make the
" experiment: he who for another purpofe (which was,
" that he might fhew the extreme velocity, with which
" the electrical power communicated itfelf, from one ex-
" tremity of a conductor to the other, however great its
" length) extended infulated iron wires to more than two
" miles in length; and to whom, on account of thefe very
" experiments, MUSCHENBROEK took occafion to addrefs
" himfelf as follows : *Magnificentiffimis tuis experimentis*
" *fuperafti conatus omnium.*   See a paper in *Opere Scelti* di
" Milano date Como 20 Aug. 1778."

Mr.

Mr. WATSON conducted some other experiments, with so much sagacity and address,
relating to the impracticability of transf-
mitting odours, and the power of purgatives
through glafs; and thofe relating to the
exhibition of what was called the " Glory
" round the Head," or the " Beatification,"
boafted to have been done by fome philofophers on the continent; that he procured,
at length, an acknowledgment from Mr.
BOSE, of what he called " an Embellifh-
" ment," in conducting the experiments;
a procedure totally incompatible with the
true fpirit of a philofopher !

Mr. WATSON's firft papers on the fubject
of Electricity, were addreffed, in three letters, to *Martin* FOLKES, Efq; Prefident of
the Royal Society, dated in March, April,
and October 1745; and were publifhed in
the *Philofophical Tranfactions* (z), under the
title of " Experiments and Obfervations
" tending to illuftrate the Nature and Pro-
" perties of Electricity." Thefe were followed in the beginning of the next year

(z) *Phil. Tranf.* Vol. xliii. p. 481—501. and Vol.
xliv. p. 695—704.

(1746)

(1746) by "Farther Experiments, &c. (*a*);" and thefe by " A Sequel to the Experi- " ments, &c,"

Thefe tracts were collected, and fepa- rately publifhed in octavo, and reached to a third or fourth edition, They were of fo interefting a nature, that they gave him the lead, as it were, in this branch of phi- lofophy; and were not only the means of raifing him to a high degree of eftimation at home, but of extending his fame throughout all *Europe.* His houfe became the refort of the moft ingenious and illuf- trious experimental philofophers that *Eng- land* could boaft.

Several of the nobility attended on thefe occafions; and his prefent Majefty GEORGE III. when Prince of *Wales,* honoured him with his prefence. In fact, there needs no greater confirmation of his merit, at that early time, as an electrician, than the pub- lic teftimony conferred upon him by the Royal Society, which, in 1745, prefented him with Sir *Godfrey* COPLEY's medal, for his difcoveries in electricity.

(*a*) *Phil. Tranf.* Vol. xliv. p. 704—749.

After

After this mark of diftinction, Mr. WAT-
SON continued to profecute electrical ftudies
and experiments, and to write on the fubject for many years. Between the year
1745, the date of his firft paper, and the
year 1764, that of the laft, we find all thofe
papers which I have recited below (*b*).

After

(*b*) Obfervations upon fo much of Monfieur *le* Mou-
NIER the younger's Memoir, lately prefented to the
Royal Society, as relates to the Communication of the
Electric Virtue to Non-electrics. Jan. 1746-7. Vol.
xliv. p. 388—395.

A Collection of Electrical Experiments. Vol. xlv.
p. 49—92. Thefe were the firft experiments made by
Mr. WATSON to determine the velocity of electricity,
and the diftance to which its power might be carried;
made on the Thames, in July and Auguft, 1747.

Further Enquiries into the Nature and Properties of
Electricity. Jan. 1748. Ib. p. 93—120.

Experiments made to determine the abfolute Velocity
of Electricity. Oct. 1748. Ib. p. 491—6. Made at
Shooter's Hill.

A Letter from Mr. *William* WATSON, F. R. S. to the
Royal Society, declaring that he, as well as many others,
have not been able to make *Odours* pafs through Glafs, by
means of Electricity; and giving a particular Account of
Profeffor BOSE at *Wittemberg*, his Experiment of *Beatification*, or caufing a Glory to appear round a Man's
Head by Electricity. March 1. 1750. Vol. xlvi. p.
348—356.

An

After writing the laft of thefe, he was appointed, by the Royal Society, one of the Committee in 1772, to examine into the ftate of the powder-magazines at *Purfleet*; and with the Honourable Mr. CAVENDISH, Dr. FRANKLIN, and Mr. ROBERTSON, fixed

An Account of Mr. *B.* FRANKLIN's Treatife, intitled, " Experiments and Obfervations on Electricity, made at " *Philadelphia*, in *America*." June 6, 1751. Vol. xlvii. p. 202—210.

An Account of Profeffor WINKLER's Experiments relating to Odours paffing through electrifed Globes and Tubes, &c.; with an Account of fome Experiments made here with Globes and Tubes tranfmitted from *Leipfic*, by Mr. WINKLER. June 20, 1751. Ib. p. 231—240.

An Account of the Phænomena of *Electricity in Vacuo*; with fome Obfervations. Feb. 1752. Ib. p. 362—375.

A Letter concerning the Electrical Experiments in *England*, upon Thunder Clouds. Dec. 21, 1752. Ib. 567—570.

An Anfwer to Dr. LINING's Query, relating to the Death of Profeffor RICHMAN. July 4, 1754. Vol. xlviii. p. 765—772.

An Account of Abbé NOLET's Treatife concerning Electricity, extracted and tranflated from the *French*. May 17, 1753. Ib. p. 201—216.

An Account of Dr. BOHADSCH's " *Differtatio Philofophico-medico de Utilitate Electrifationis in curandis Morbis*;" printed

fixed on pointed conductors as preferable to blunt ones; and again, was of the Committee in 1778, after the experiments of Mr. WILSON in the Pantheon.

printed at *Prague* in 1751. Extracted and translated from the *Latin*. Jan. 23, 1752. Vol. xlvii. p. 345—351.

An Account of Dr. BIANCHINI's "*Recueil d'Experiences faites à Venise sur le Medicine Electrique*." March 12, 1752. Ib. p. 399—406.

An Account of a Treatise in *French*, intitled " *Lettres sur l'Electricité*;" by the Abbé NOLET. Dec. 17, 1761. Vol. lii. p. 336—343.

Suggestions concerning the preventing the Mischiefs which happen to Ships and their Masts by Lightning; in a Letter to *George* Lord ANSON, First Lord of the Admiralty. Dec. 1762. Ib. p. 629—635.

Observations on the Effects of Lightning; with an Account of the Apparatus proposed to prevent its Mischiefs to Buildings, more particularly to Powder Mills. Being Answers to certain Questions proposed by *M.* CALANDRINI, of *Geneva.* June 28, 1764. Vol. liv. p. 201—227. Including an Account of the Mischief *St. Bride's* Steeple sustained by Lightning on the 18th of June 1764.

CHAP. 51.

Account of Sir William Watſon continued — His great acquaintance with the police of the city of London — Miſcellaneous papers written by him —His traÑts on medical ſubjeÑts printed in the Philoſophical TranſaÑtions — Zoological papers—Created DoÑtor of Phyſic by the Univerſities of Halle and of Wittemberg—His experiments on inoculation—His medical writings in the London Medical Obſervations — Conſtituted one of the Vice-Preſidents of the Royal Society — EleÑted Fellow of the College of Phyſicians — Has the honour of knighthood conferred on him—His death, and charaÑter.

WATSON.

AS Mr. WATSON had conſtantly lived in *London*, he had been a curious obſerver of the wonderful increaſe and improvement of that vaſt city. He was acquainted, in no ordinary degree, with its hiſtory, and its police in general ; and had particularly attended to thoſe circumſtances that were more immediately the objeÑts of
the

the philofopher and the phyfician. This
knowledge enabled him frequently to fug-
geft ufeful hints; one of which highly deferves to be mentioned, as it refpects an
object of great importance to the public.

In the hard winter of 1756, he wrote
" Some Obfervations on preventing the
" freezing of Water in the Leaden Pipes
" of the City of *London;*" occafioned by
the injudicious and ineffectual method,
practifed frequently, of ftrewing dung in
the ftreets over the pipes. Thefe were
printed in the *Gentleman's Magazine* (*a*) for
January 1757, p. 6. in which is pointed

(*a*) The method was fimply by means of two additional brafs cocks. One to be inferted into the leaden
pipe, two feet before it comes into the air, guarded by a
wooden cafe, filled up with horfe litter, and reaching near
to the furface of the ground, and covered over, even with
the ground, by a brick or ftone. This is to ferve as a
ftop-cock, and to be turned by the help of an iron key.
The other cock is to be faftened to the leaden pipe in
the open air, in any part of its length, provided it be
fomewhat below the level of the ftop-cock. This is inferted fimply to empty the leaden pipe of all its water,
after it has been turned off by the ftop-cock. From the
defcription of this apparatus, the method of ufing it is obvious.

out

out a fuccefsful method of effecting the
purpofe, which he had himfelf employed
in the fevere winter of 1739-40. Other
inftances, befides this, occur, of his attention to whatever might advance the welfare of the public. So early as the year
1742, he had laid before the Royal Society " Some Obfervations upon Mr. *Sut-*
" *ton*'s Invention to extract the foul and
" ftinking Air from the Well and other
" Parts of Ships: With critical Remarks
" upon the Ufe of Windfails." In which
he fuggefts feveral improvements in that
ufeful invention (*b*).

In 1753, he publifhed Mr. *Appleby*'s
Procefs for rendering Sea-water frefh (*c*).

In 1768, an Account of Mr. *Charles*
Miller's Experiments on the fowing of
Wheat, and dividing the Root; by which
means were produced, in one year, from
one grain, 21,109 ears, which yielded three
pecks and three quarters of clean corn,
weighing forty-feven pounds feven ounces;

(*b*) *Phil. Tranf.* Vol. xlii. p. 62—70.
(*c*) Vol. xlviii. p. 69.

and the number of grains, calculated by the number in one ounce, might be 576,840 (*d*). It is to be feared that this method can fcarcely be reduced to advantageous practice on a large and agricultural plan.

In the fame year, an Account of the Oil extracted from the *American* Earth-nut, or, more properly, *Ground Peafe* (*e*). This plant, like a few others of the fame clafs, has the fingular property of protruding its feed-veffel into the ground, where it ripens the fruit; hence it is named by RAY, *Arachis Hypogaios*. The oil of this pulfe is fo mild and well tafted, and withal fo eafily procured, that it might bid fair to fuperfede that of olives, or even oil of almonds. It is cultivated in *North Carolina*, and might advantageoufly be raifed in the Sugar Iflands (*f*).

As from the earlieft times of the Royal Society, it had been cuftomary to requeft of fome member, properly qualified from his knowledge of the fubject, to review, and

(*d*) *Phil. Tranf.* Vol. lviii. p. 203.
(*e*) Arachis hypogæa *Lin. Spec. Plant.* p. 1040.
(*f*) *Phil. Tranf.* Vol. lix. p. 379—383.

lay

lay before that body at their uſual meetings, any ſuch extracts from the numerous publications which were ſent to the Society, relating to diſcoveries in philoſophy and the arts, as promiſed to be of general utility, that they might be recorded in the *Philoſophical Tranſactions,* this office did not unfrequently fall upon Mr. WATSON. We find ſeveral papers of this nature bearing his name. Beſides thoſe which I have enumerated, relating to natural hiſtory ſtrictly, and electricity, are the following, as recited below (*g*).

Of his productions which have a more immediate reference to phyſic, the firſt was publiſhed in the *Philoſophical Tranſactions,* N° 459. " A Caſe wherein Part of the " Lungs were coughed up." And in the ſucceeding Number, " An Obſervation re-

(*g*) An Account of a Book, intitled " *De quamplurimis Phoſphoris nunc* primum *detectis Commentarius* Auctore *Jac. Barthol. Beccaria.*" 4°. *Bolog.* 1744. Feb. 1746. Vol. xliv. p. 81—91.

An Account of a Treatiſe in *Latin,* dedicated to the Royal Society, intitled " *Commentatio de Prærogativa Thermarum Carolinarum in diſſolvendo Calculo Vericæ præ Aqua Calcis vivæ. Auct. G. C. Springsfeld.*" Vol. xlix. p. 895 —906.

" lating

" lating to Hydatides voided *per Vagi-*
" *nem* (*h*)."

In 1744, an Account, and Analyfis, of a
Stone, which, when firft taken out of the
ftomach of a coach-horfe, weighed three
pounds two ounces avoirdupois weight, and
meafured feventeen inches by fixteen.

On examination, it appeared to be not fo
much a concretion of the kind called *Ega-
gropila*, as of the *bezoardic* texture (*i*). Mr.
WATSON had afterwards an opportunity of
exhibiting to the Society a *Calculus*, taken
from the belly of a mare, which weighed
fifteen pounds twelve ounces. Even this,
however, was exceeded by one from a dray-
horfe belonging to Sir *Henry Hicks* at *Deptford*, which weighed nineteen pounds, exclufive of fome of the cruft broken off (*k*).

In 1749, he laid before the Royal Society, " An Account of the *Vomito Prieto* of
" *Carthagena*," called on the fpot *La Chap-
petonade*. This was extracted from Don
ULLOA's Voyage to *South America*, juft

(*h*) *Phil. Tranf.* Vol. xliii. p. 623. and p. 711.
(*i*) Ib. p. 268.
(*k*) Vol. xlviii. p. 800.

                                        then

then publifhed at *Madrid* (*l*). This difeafe
is defcribed by SAUVAGES under the name
of *Vomitus rabiofus*.

In the fame volume, " Cafes of the *Fœ-*
" *tus in Utero* being differently affected by
" the Small-pox." In one of thefe, a female child was born with evident marks of
the fmall-pox upon her, and was not fuf-
ceptible of the difeafe when inoculated at
four years old with her brother, who paffed
through it very favourably. The girl grew
pale, and loft her appetite ; but her indif-
pofition wore off in two or three days. The
other is the cafe of a lady, who had the
fmall-pox to a great degree when feven
months gone with child, which was at the
fame period of pregnancy under which the
mother of the above-mentioned child paffed
through the diftemper. The offspring of
this lady, however, went through the difeafe in the natural way, at the age of four
or five years (*m*).

All who were acquainted with the ex-

(*l*) *Phil. Tranf.* Vol. xlvi. p. 134.
(*m*) Ib. p. 235.

Y 3   tent

tent of Mr. WATSON's knowledge in the
practice of physic, in natural history, and
experimental philosophy, were not surprised
to see him rise into the higher line of his
profession. This event took place in 1757,
previous to which he had been chosen a
member of the Royal Academy of *Madrid*;
and he was created doctor of physic by the
University of *Halle*, under a diploma, bearing date September the 6th. The same honour was conferred upon him by that of *Wittemberg* about the same time. Soon after
which he was disfranchised from the Company of Apothecaries. He became a licentiate of the College of Physicians in
1759.

This alteration in his circumstances and
prospects, hazardous as it might be considered by some, occasioned no diminution in
his emoluments, but far the contrary. He
had before this time removed from *Alders-
gate Street* to *Lincoln's Inn Fields*, where he
lived the remainder of his days ; and now
he found himself at greater liberty to pursue
his studies, and carry on at more leisure the
extensive literary connnexion in which he

was

was engaged, both at home and abroad. He kept up a clofe correfpondence with Dr. HUXHAM for many years. We find among his correfpondents abroad, the names of M. PEYSSONNEL, CLAIRAUT, BOSE, the Abbé NOLLET, M. ALLEMAND, M. JUS-SIEU, and many others, as may be feen from the letters communicated by him to the Royal Society.

In October 1762, Dr. WATSON was chofen one of the phyficians to the Found-ling Hofpital, which office he held during the remainder of his life.

We find alfo two zoological articles laid before the Royal Society by Dr. WATSON. The firft of thefe relates to the infect called the *Vegetable Fly,* which had impofed on the credulity of many, under the idea of its being an infect flying about with a vegetable growing on its back: whereas in fact it was nothing more than a fungus of the *Clavaria* genus, growing from the dead nymph of a *Cicada,* as well as from any other putrid animal fubftances (*n*). The firft author who feems to have counte-

(*n*) *Phil. Tranf.* Vol. liii. p. 271. tab. 23.

nanced

nanced this error, was Father TORRUBIA, in his " *Apparato para la Hiftoria naturali Efpanola*," printed at *Madrid*. Fol. 1754. He defcribes and figures a prickly plant, vegetating from a dead wafp. Both thefe productions are figured by Mr. EDWARDS, in the third part of his " Gleanings," tab. 335, 336.

The fecond paper is a Defcription, accompanied by a large engraving of the American Armadillo, called *Dafypus no-vemcinctus* by LINNÆUS, the nine-banded Armadillo (*o*).

In 1758, was printed part of a letter to Dr. HUXHAM, being an account of fome extraordinary effects arifing from convulfions, in a young lady, which ended in a deprivation of fpeech, and temporary blindnefs. Thefe fymptoms lafted fourteen months, and were at laft fuddenly removed after fhe had heated herfelf by four hours dancing (*p*).

" Some Obfervations relating to the *Lyn-* " *curium* of the Antients;" tending to prove

(*o*) *Phil. Tranf.* Vol. liv. p. 57. t. 7.
(*p*) Ib. Vol. l. p. 743.

that

that it was the *Tourmalin* of the moderns (*q*).

In 1762, a Letter to Dr. HUXHAM, containing fome Remarks on the *Influenza* of that year, and on the Dyfentery which fucceeded it (*r*).

Obfervations upon the Effects of Electricity, applied to a *Tetanus,* or mufcular rigidity, of four months continuance. For the firft three weeks the ftiffnefs was confined to the jaw, but afterwards extended to a total rigidity of the fpine. Electrization was continued for ten weeks with a fenfible advantage, and the girl was wholly reftored to health (*s*).

In 1764, Dr. WATSON laid before the Royal Society " An Account of what ap-
" peared on opening the Body of an Afth-
" matic Perfon." This was a young man, aged twenty-eight, who died after being afflicted with an afthma only two months. The lungs were found in an extraordinarily emphyfematous ftate, and the pulmonary

(*q*) *Phil. Tranf.* Vol. li. p. 394.
(*r*) Ib. Vol. lii. p. 646.
(*s*) Ib. Vol. liii. p. 10—26.

vein

vein varicofe in a great degree. A forenefs
of the cheft, fucceeded by a cough and a
fhortnefs of breath, had in this young man's
cafe immediately fucceeded a violent and
long-continued vomiting; to which caufe
Dr. WATSON was inclined to attribute the
origin of this difeafe (*t*).

Part of a Letter to Dr. HUXHAM, giving
fome account of the late cold weather, dated
London, Feb. 14, 1767. By this it appears
that the thermometer in *London* ftood, when
at the loweft, on the 19th, at eight in the
morning, at $15''\frac{1}{2}$: and on the fame day, at
Norwich, it was obferved as low as feven
degrees (*u*).

In 1768, Dr. WATSON publifhed " An
" Account of a Series of Experiments, in-
" ftituted with a view of afcertaining the
" moft fuccefsful Method of inoculating
" the Small-pox." 8°. Thefe experiments
were defigned to prove whether there
was any *fpecific* virtue in preparatory medicines: whether the difeafe was more favourable when the matter was taken from

(*t*) *Phil. Tranf.* Vol. liv. p. 239—245.
(*u*) Ib. Vol. lvii. p. 443.

the

the natural, or the artificial pock: and, whether the crude lymph, or the highly-concoded matter, produced different effects. The refult was, what fucceeding and ample experience has confirmed, that after due abftinence from animal food, and heating liquors, it is of fmall importance what kind of variolous matter is ufed; and that no preparatory fpecifics are to be regarded.

Of Dr. WATSON's papers on medical fubjeds, printed in other publications, it will be unneceffary to give a detailed account; as they are well known to medical practitioners in general. Neverthelefs, that the lift of his produdions may be complete, I fhall recite them briefly.

" An Account of the good Effeds of *Magnefia* in fevere Vomitings (*x*)."

" Obfervations on the *Hydrocephalus internus* (*y*)."

" An Account of the Putrid Meafles, as

(*x*) *London Medical Obfervations*, Vol. iii. p. 335—340.

(*y*) Ibid. Vol. iv. p. 78—88.

5 " they

" they were obferved in *London* in the years
" 1763 and 1768 (*z*).

" An Appendix to the Paper on the *Hy-*
" *drocephalus internus* (*a*).

This difeafe, on which Dr. WHYTT,
Dr. WATSON, and others, have lately written in fo inftructive a manner, deferves to
be accurately noticed, and the knowledge
of it ftrongly inculcated; as, in the country at leaft, it is not unfrequently mifta-
ken, and treated as a putrid and comatofe
fever.

As Dr. WATSON lived in intimacy with
the moft illuftrious and learned Fellows
of the Royal Society; fo he was himfelf
one of its moft active members, and ever
zealous in promoting the ends of that inftitution. For many years he was a frequent member of the council; and, during
the prefidentfhip of Sir *John* PRINGLE, was
elected one of the vice-prefidents; which
honourable office he continued to fill to the

(*z*) *London Medical Obfervations*, Vol. iv. p. 132—
155.

(*a*) Ibid. p. 321—329.

end

end of his days. He was a moft conftant attendant on the public meetings of the Society; and on the private affociations of its members, efpecially on that formerly held every Thurfday, at the Mitre, in *Fleet Street*, and now at the Crown and Anchor Tavern, in the *Strand*.

In 1784, Dr. WATSON was chofen a Fellow of the Royal College of Phyficians; and made one of the Elects. In the fucceeding year, he communicated to the College, " An Account of a difeafe occa- " fioned by tranfplanting a Tooth." This was inferted in the *Medical Tranfactions*; and this, I believe, was the laft paper he wrote (*b*).

In 1786, he had the honour of knighthood conferred upon him; being one of the body deputed by the College to congratulate his Majefty on his efcape from affaffination.

In general, Sir *William* WATSON enjoyed a firm ftate of health. It was fometimes interrupted by fits of the gout; but thefe feldom confined him long to the houfe. In

(*b*) P. 325—338.

the

the year 1786, the decline of his health was
very vifible to his friends, and his ftrength
was greatly diminifhed, together with much
of that vivacity which fo ftrongly marked
his character.   He died May 10, 1787.

Sir *William* WATSON had a natural activity both of mind and body, that never
allowed him to be indolent in the flighteft
degree.   He was a moft exact œconomift
of his time, and throughout life a very early
rifer, being up ufually in fummer at fix
o'clock, and frequently fooner; thus fecur-
ing to himfelf daily two or three uninterrupted hours for ftudy.   In his younger
days, thefe early hours, as I have before
obferved, were frequently given up to the
purpofes of fimpling; but, in riper years,
they were devoted to ftudy.   He read much
and carefully; and his ardent and unremit-
ting defire to be acquainted with the progrefs of all thofe fciences which were his
objects, joined to a vigorous and retentive
memory, enabled him to treafure up a vaft
ftock of knowledge.   What he thus acquired, he freely difpenfed.   His mode of
conveying information was clear, forcible,

                                            and

and energetic, and juftified the encomium beftowed upon him by a learned foreigner, in a letter to a correfpondent (*c*).

His attention, however, was by no means confined to the fubjects of his own profef- fion, or thofe of philofophy at large. He was a careful obferver of men, and of the manners of the age; and the extraordinary endowment of his memory had furnifhed him with a great variety of interefting and entertaining anecdotes, concerning the cha- racters and circumftances of his time (*d*).

On all fubjects, his liberal and commu- nicative difpofition, and his courteous be- haviour, encouraged enquiry; and thofe who fought for information from him, fel-

(*c*) WATSONIUS *Botanicus et Phyficus clarus eft et perfpicax homo, itidemque humaniffimus.* M. Meckel, of *Berlin,* in Epiftolis ad HALLERUM *datis.*

(*d*) It is to Sir *William* WATSON that we owe the pre- fervation of an anecdote, which tends further to illuftrate the character, and exalt the fincerity and integrity of the excellent Mr. ADDISON. It is inferted in the *Addenda* to his Life, in the third volume of the *Biographia Britan- nica.* Dr. KIPPIS alfo acknowledges himfelf the moft indebted to him for the materials of the life of the late *Henry* BAKER, Efq.

dom

dom departed without it. In his epistolary
correspondence he was copious and precise;
and such as enjoyed the privilege and pleasure of it, experienced in his punctuality
another qualification which greatly enhanced its value.

Some of the first of Sir *William* WAT-
SON's papers in the *Philosophical Transactions*, evince his early proficiency in the
science of Botany, and especially his acquaintance with the *English* species : nor
was he less skilled in exotics in his riper
years. That he was very soon considered
on the continent as highly respectable in this
light, is manifest from his having been one
of the few in *England*, whom Mr. CLIFFORD
gratified with a copy of the *Hortus Cliffortianus*; a work, at its first publication,
only attainable by those whose studies and
acquirements in the subject of it, entitled
them to receive it from the munificence of
Mr. CLIFFORD himself. In fact, all learned foreigners, of the same bias in their studies, brought letters of recommendation to
him; and, on their return, failed not, both
in their correspondence and in their writings,

tings, to bear honourable teſtimony to his learning and abilities.

Sir *William* WATSON had learned to know plants by the ſyſtem and nomencla- ture of RAY, when *trivial* names were unknown; and he was ſo ſingularly happy in a tenacious memory, as to be able to repeat, with wonderful promptitude, the long names which had been in uſe from the times of BAUHINE, GERARD, and PARKINSON; a taſk from which botaniſts are relieved, by the introduction of the *Linnæan trivial* epithets. He lived to ſee the ſyſtem of his much-honoured country- man give way to that of the *Swede*, which began to take place in *England* about this period; and with which alſo he made him- ſelf acquainted. His knowledge of plants, and the hiſtory of them in the various au- thors, was ſo eminently extenſive, that his opinion was frequently appealed to as deci- ſive on the ſubject; and by ſome of his in- timate friends he was uſually called " The " living Lexicon of Botany." Had it been the lot of Sir *William* WATSON to have been devoted to Botany as an official employment; or had the more important avocations of his profeffion allowed a further indulgence to his favourite bias, fuch an union of natural endowments and acquired knowledge as he poffeffed, muft have placed him very high among the naturalifts of this age.

It remains for me to do juftice to the worth of Sir *William* WATSON as a phyfician, and as a member of fociety. But as thefe parts of his character have been already delineated with great truth and difcrimination by my much-refpected friend Dr. GARTHSHORE, I fhall conclude this account by fome extracts from the Memorial read by him to a fociety of phyficians, of which Sir *William* had been the prefident.

" As a phyfician, his humanity, affiduity,
" and caution, were eminently confpicuous;
" and his exact obfervance of the duties of
" focial politenefs muft ever be remember-
" ed with pleafure by all thofe who en-
" joyed the happinefs of his acquaintance.
" The fmile of benignity was always dif-
" played on his countenance; he invaria-
" bly

" bly continued the general, the ready, and
" the obliging friend of mankind; he was
" refpectful to the elder and fuperior, en-
" couraging to the younger, and pleafant
" and eafy to all with whom he had any
" intercourfe. The fame affability and good
" humour which adorned his character in
" public life, were preferved alfo in the bo-
" fom of his family, and endeared him to
" thofe who were more immediately around
" him. He was fcarcely ever out of tem-
" per, was always benignant and kind to
" his friends and relations—and, it would
" be injurious to his memory not to men-
" tion an anecdote which equally difplays
" his humanity, and the warmth with
" which he interefted himfelf in the cafes
" of his patients—Not many years before
" his death, he was waked fuddenly one
" morning very early by his fervant, who
" came to inform him that his houfe had
" been broken open, and that his plate
" (which was of confiderable value) was
" ftolen—" Is that all?" faid he, coolly—
" I was afraid you had brought me fome
" alarming meffage from Mr. ——, con-

" cerning

" cerning whofe dangerous fituation I have
" been very uneafy all night (*e*)."

(*e*) In 1759, Mr. MILLER paid Dr. WATSON the
tribute of calling a new genus in the *Triandrous* clafs
after his name ; two fpecies of which he has figured in
the " Cuts adapted to the Gardener's Dictionary," tab.
276. and tab. 297. fig. 2. It proved that Dr. TREW had
before given the name of *Meriana* to the firft of thefe ;
and LINNÆUS found himfelf obliged by the rules of his
fyftem, to reduce thefe two fpecies to his genus *Antholyza*,
already eftablifhed in the *Species Plantarum*; thus finking
the generic term of *Watfonia*, and retaining TREW's as
a *trivial* name to the plant of tab. 276. It is to be regretted that, in juftice to Dr. WATSON, who had de-
ferved fo eminently well of the fcience, that LINNÆUS
did not at leaft name the leffer fpecies, tab. 297. 2. of
MILLER, *Antholyza Watfonia*, inftead of *A. Merianella*.

C H A P. 52.

Linnæus—*vifits* England—*Cool reception of him by Sir Hans* Sloane — Dillenius *fenfible of his merit*; *but indifpofed to receive the fexual fyf-tem* — *Botany at this juncture in a languid ftate in* England — Linnæus's *writings diffufed in* England *about the year* 1740 — Grufberg's Flora Anglica — Brown's *Jamaica Plants* — Stillingfleet's *Tracts* — Lee's *Introduction* — Hill's Flora Britannica — Hudfon's Flora Anglica — *Dr.* Solander — Linnæus's *fyftem adopted in the public lectures at* Cambridge *and at* Edinburgh—*and, finally, received and eftablifhed in* England.

L I N N Æ U S.

AS I am now arrived beyond the pe-riod, when the name of LINNÆUS began to be celebrated throughout *Europe*, it will be neceffary to recur to the circum-ftances of his vifit to this country, that the introduction and full eftablifhment of his fyftem in this kingdom, may be better illuftrated. Here, had his reception been

more

more encouraging to his wiſhes, it has been
ſaid, he was diſpoſed to have taken up his
reſidence. He had been ſome time in *Holland*, under the patronage, and in the houſe,
of Mr. CLIFFORD. He had taken his degree of doctor in phyſic. He had gained
the eſteem of BOERHAAVE, and from him
brought letters of recommendation to the
literati of *England*.

The fame of Sir *Hans* SLOANE and his
Muſeum, and the eſteem in which LIN-
NÆUS held the character of DILLENIUS,
added to the deſire of inſpecting the *She-
rardian Pinax*, were among the moſt powerful motives that induced the *Swede* to viſit
England. This event took place in the
ſpring of 1736. I am only able to aſcer-
tain the ſeaſon of the year, from being informed of the pleaſure he expreſſed, in
meeting in the fields with thoſe productions of *England*, that are not ſpontaneouſly
growing in *Sweden*. His delight particularly, in ſeeing under the hedges the *Hya-
cinth* in full flower, can only be conceived
by thoſe who poſſeſs ſome ſhare of that botanical ardour which he poſſeſſed.

§   At

At this time, the fexual fyftem exifted only in its outline. Enough of it, however, was manifefted in the *Florula Lapponica*, printed in the *Acta Upfalienfia*, for the years 1732 and 1733, and in the firft fketch of the *Syftema*, in 1735, to exhibit its novelty. I know not that the *Fundamenta Botanica*, the *Bibliotheca*, and the *Mufa Cliffortiana*, although they bear date in 1736, had reached *England* before the author: yet, notwithftanding the warm recommendation of BOERHAAVE, Sir *Hans* SLOANE, confidered at that time as the *Mecænas* of Botany in this ifland, gave the author, and his fyftem, an unfavourable reception. At the age of feventy-fix, we need not be furprifed that the veteran fhould not feel difpofed to learn a new fyftem, from a young man, whom he could not but confider as an adventurer, both in fortune, in fame, and in fcience. SLOANE, moreover, had never paid fufficient attention to the improvement of fcience in the conftruction of generical characters; and this circumftance, probably, fet him at a farther diftance from embracing the fyftem of

Z 4   LINNÆUS,

LINNÆUS, which exhibited an arrangement so widely different from the undefined assemblage of the History of *Jamaica.*

It must not however be understood, that Sir *Hans* SLOANE remained insensible to the genius and accomplishments of LINNÆUS: on the contrary, when he afterwards sent him his *Flora Lapponica,* Sir *Hans* SLOANE wrote him a letter, bearing date Dec. 20, 1737, expressive of the great pleasure he received in the perusal of it; exhorting him to elucidate the remaining parts of the natural history of his country, on the same plan.

DILLENIUS was highly sensible of his merit, and gave him the most polite reception. But that he who had been so long versed in the systems of TOURNEFORT and RAY, and after having given improvement to the latter, by which he had deserved and received the applause, not of *England* alone, but of all *Europe,* should abandon that system, to embrace the hitherto uncountenanced novelties of LINNÆUS, could not reasonably be expected.

The

The journey into *England* however, was, on the whole, highly gratifying to LIN- NÆUS. He beheld with aftonifhment the collections of SLOANE, and, with rapture, the *Herbaria* of PETIVER, PLUKENET, BU- DELLE, and of many others there repofit- ed, whofe names were familiar to him. At *Oxford* he infpected, with no lefs fatisfac- tion, the *Pinax* of SHERARD, which he had eagerly wifhed to fee publifhed, and of which DILLENIUS had compleated about a fourth part. But an undertaking of that nature and extent, after the death of the firft projector of it, demanded a patronage and an expence, not eafily obtained.

About the time LINNÆUS made his tour into this country, indigenous botany was on the whole in a languifhing ftate. It no longer felt that degree of fupport, which the SHERARDS, and Sir HANS, had afford- ed it. The Conful was dead; and the de- clining years of Dr. *James* SHERARD, and of Sir *Hans* SLOANE, began to withdraw them from the buftle, and almoft from the bufinefs, of life. After the publication of RAY's *Synopfis* by DILLENIUS, in 1724,

no

no work of magnitude on the *Englifh* botany, except the *Hiftoria Mufcorum*, in 1741,
took place for many years; not that there
were wanting feveral individuals, who were
eminent for their knowledge of indigenous
botany, and zealous in propagating it: as
inftances, I refer to the names of WATSON,
COLLINSON, MILLER, and BLACKSTONE.
The arrival however of LINNÆUS in *England*, and the confequent promulgation of
his method, excited that curiofity which
novelty will ever attract, and, although his
fyftem might be but little relifhed at the
inftant, by the *Englifh* naturalifts in general, there were yet a few into whofe minds
his doctrines filently infinuated themfelves,
and gained approbation.

In the year 1737, the next after LIN-
NÆUS left *England*, he publifhed the *Genera Plantarum*, which compleatly un-
folded the fexual fyftem, as far as related to
claffical and generical characters; and in
the fame year exemplified it in the fpecies,
by the *Flora Lapponica*, and the *Hortus
Cliffortianus*. At the fame time, anxious
as it fhould feem above all, to gain the approbation

probation of Dillenius, he dedicated to him the *Critica Botanica*; in which he explains his reasons for the change of names, and for the establishment of new distinctions, both of which, he well knew, would be considered as dangerous innovations.

These volumes soon found their way into the libraries of the curious in *England*; though the *Hortus Cliffortianus* was, at first, only dispersed through the munificence of Mr. Clifford. The simplicity of the classical characters as the basis, the uniformity of the generical notes, confined wholly to the parts of fructification, and that precision which marked the specific distinctions, advantages, of which all foregoing systems were destitute, soon commanded the assent of the unprejudiced; and an interval of a few years, gave Linnæus's method a decided superiority with *English* botanists.

After the establishment of Linnæus in the professorship in the year 1741, the publication of the *Theses*, afterwards, in a collected form, called the *Amœnitates Academicæ*, commenced, and, in less than ten

years,

years, two volumes had been published.
These tracts, by the variety of useful and
entertaining knowledge, with which they
abound, equally extended and augmented
the reputation of LINNÆUS. They convinced his opposers, that his knowledge
was not bounded by mere nomenclature,
and systematic arrangement, as was re-
proachfully objected.

CONCLUSION.

In *England*, Dr. MARTYN, in his *Virgil*, published in 1740; DILLENIUS, in his
Historia Muscorum, 1741 ; and BLACK-
STONE, in his *Specimen Botanicum*, 1746,
had referred to the writings of LINNÆUS;
and occasionally his name had been mentioned in the *Philosophical Transactions*, and
other periodical works : but, as yet, no
translation of any part of his writings, or
any publication on his plan, had been made
in this country, until, in 1754, a *Swedish*
pupil of the *Upsal* school arranged, by the
generic and trivial names only, all the plants
of RAY's *Synopsis*, according to the system
of his master. This little tract was immediately

mediately tranfmitted to the Royal Society, and excited much attention among thofe profeffed ftudents, and lovers of *Englifh* botany, who obtained the perufal of it.

In 1756, Dr. BROWNE claffed all his *Jamaica* plants, amounting to 1200 fpecies, in the fame method. The drawings having been made by EHRET, had the advantage of feparate delineations of the flower and fruit.

In 1759, Mr. STILLINGFLEET pub-lifhed a Tranflation of feveral Tracts from the *Amœnitates* ; and, by his own valuable additions, his inftructive Preface, the judi-cious and learned notes interfperfed through-out the book, by his own " Calendar of " Flora," confirming and illuftrating that of the *Swede*, greatly conduced to exalt the reputation of LINNÆUS in *England*. Of this learned and excellent man, the reader will find fome memoirs in the *Gentleman's Magazine* for 1776, which were afterwards incorporated into " Anecdotes of Mr. *Bow-* " *yer*" (fee p. 300), and into the *Biogra-phia Dramatica*, 2d edition, 1782.

The next year, Mr. LEE, by his Tranf-lation

lation of the *Elements* of the Sexual Syftem,
much contributed to facilitate the knowledge, and extend the progrefs and popu-
larity of it, among the lefs learned of his
countrymen, or fuch as were unable to recur to the *Fundamenta*, or *Philofophia Botanica* of the author.

At this juncture, it is material among
thofe circumftances which accelerated the
progrefs of the new fyftem, to mention the
arrival of the late much-lamented Dr. So-
LANDER, who came into *England* on the
firft of July, 1760. His name, and the
connexion he was known to bear as the
favourite pupil of his great mafter, had of
themfelves fome fhare in exciting a curiofity which led to information ; whilft his
perfect acquaintance with the whole fcheme
enabled him to explain its minuteft parts,
and elucidate all thofe obfcurities with
which, on a fuperficial view it was thought
to be enveloped. I add to this, that the
urbanity of his manners, and his readinefs
to afford every affiftance in his power,
joined to that clearnefs and energy with
which he effected it, not only brought
conviction

conviction of its excellence in those who were inclined to receive it, but conciliated the minds, and dispelled the prejudices, of many who had been averse to it.

By all these preliminary advances, the learned were prepared to see the *English* botany modelled according to the rules of the *Linnæan* school. Dr. HILL seized the first opportunity of attempting it, in his *Flora Britannica,* 1760; but it was executed in a manner so unworthy of his abilities, that his work can have no claim to the merit of having answered the occasion: and thus the credit of the atchievement fell to the lot of Mr. *William* HUDSON, F. R. S. who, to an extensive knowledge of *English* plants, acquired by an attention to nature, had, by his residence in the *British Museum,* all the auxiliary resources that could favour his design: access particularly to the *Herbaria* of almost all the assistants of RAY and DILLENIUS, mentioned in the *Synopsis,* gave him the opportunity of comparing the individual specimens of that work with his own; and thus enabled him to dispel a multitude of doubts and uncertainties, in which,

which, otherwife, his application of the *fy-nonyma* might have been involved.

The fexual fyftem was received nearly about the fame time in the univerfities of *Britain*; being publicly taught by Mr. Profeffor MARTYN, at *Cambridge,* and by Dr. HOPE, at *Edinburgh*. The adoption of it by thefe learned Profeffors, I confider, therefore, as the æra of the eftablifhment of the *Linnæan* fyftem in *Britain*—a fyftem, which, if I may be allowed the expreffion, had given the author of it a literary domi-nion over the vegetable kingdom; which, in the rapidity of its extenfion, and the ftrength of its influence, had not perhaps been paralleled in the annals of fcience.

I N D E X.

INDEX.

A.

I N D E X.

5

       Botany,

C.

       Cassia

DEERING,

INDEX.

I N D E X.

F.

G.

H.

HASSELQUIST,

I N D E X.

I.

INDEX.

LYTE,

I N D E X.

N.

I N D E X.

I N D E X.

INDEX.

RAY,

I N D E X.

RAY,

                                       " Select

Smith,

INDEX.

T.

I N D E X.

END OF THE SECOND VOLUME.

VOL. I.

Page 9. line 9. *for* Deus *read* Dens.
— 112. — *ult. after* 1612 *add* Quere?
— 121. — *ult. for* Dutch *r.* German.
— 177. — 14. — LINNÆUS *r.* RUPPIUS.
— 248. — 5. — Allorfinarum *r.* Altorfi-
                    narum.
— 266. — 8. — CAMELL *r.* KAMEL.
— 301. — 19. — BRUYNER *r.* BRUNYER.
— 335. — 14. — Polifh *r.* Bohemian.
— 359. — 4. — olympicum *r.* calycinum.

VOL. II.

— 57. — 5. *and elsewhere, for* KREIG *read*
                    KRIEG.
— 102. — 21. *after* 1773 *add* 1779, *and* Wil-
                    liam WHEELER 1780.
— 214. — 14. *for* Phrenanthes *r.* Prenanthes.
— 227. — 15. *of the note, for* Petals *r.* Invo-
                    lucra.
— 231. — 10. *and elsewhere, for* HOUSTON *r.*
                    HOUSTOUN.
— 332. — 21. *for* elected *r.* appointed.
— 345. — 5. — BUDELLE *r.* BUDDLE.

REMARKS.

VOL. I. page 91. line 8. Note.

I am informed by the favour of Mr. *Dryander*, that even the merit of this improvement, cannot be afcribed to *Lyte*, for that it exifts in the Tranflation made by *Clufius*.

Page 57. line 18.

There is reafon to doubt whether even this MS. was in England at this time ; fince the Norfolk Collection was *chiefly* made by *Thomas* Earl of *Arundel* and *Surrey*, in the beginning of the laft century.—Mr. *Dryander*.

VOL. II. page 28. line 21.

Part of *Plukenet*'s Herbarium was in the poffeffion of the late *Philip Carteret Webb*, Efq; and was difpofed of at the Sale of his Books.—*ib.*

Page 150. line 1.

I am informed by Mr. *Dryander*, that thofe Manufcripts confift of Dr. *Sherard*'s Literary Correfpondence. Thefe Letters are bound in five volumes *folio.*

Page 182. line 1. of the Note.

The original Drawings of the Plates in the *Hiftoria Mufcorum*, are in Sir *Jofeph Banks*'s Library. They were bought at the Sale of Drawings belonging to the late *Robert More*, Efq; of *Shropfhire.*—*ib.*

Printed in the United States
By Bookmasters